Mark Anthony Benvenuto
Industrial Biotechnology

T0291029

Also of interest

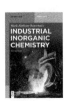

Industrial Inorganic Chemistry
Benvenuto, 2024
ISBN 978-3-11-132944-4, e-ISBN (PDF) 978-3-11-132951-2

Industrial Chemistry
Benvenuto, 2023
ISBN 978-3-11-067106-3, e-ISBN (PDF) 978-3-11-067109-4

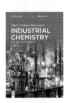

Industrial Chemistry.
For Advanced Students
Benvenuto, 2023
ISBN 978-3-11-077874-8, e-ISBN (PDF) 978-3-11-077876-2

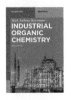

Industrial Organic Chemistry
Benvenuto, 2024
ISBN 978-3-11-132991-8, e-ISBN (PDF) 978-3-11-133035-8

Chemistry for Biomass Utilization
Alén, 2023
ISBN 978-3-11-060834-2, e-ISBN (PDF) 978-3-11-060836-6

Mark Anthony Benvenuto

Industrial Biotechnology

2nd Edition

DE GRUYTER

Author
Prof. Mark A. Benvenuto
University of Detroit Mercy
Department of Chemistry & Biochemistry
4001 W. McNichols Rd.
Detroit, MI 48221-3038
USA
benvenma@udmercy.edu

ISBN 978-3-11-132993-2
e-ISBN (PDF) 978-3-11-133025-9
e-ISBN (EPUB) 978-3-11-133082-2

Library of Congress Control Number: 2024938444

Bibliographic information published by the Deutsche Nationalbibliothek
The Deutsche Nationalbibliothek lists this publication in the Deutsche Nationalbibliografie;
detailed bibliographic data are available on the Internet at http://dnb.dnb.de.

© 2024 Walter de Gruyter GmbH, Berlin/Boston
Cover image: kontrast-fotodesign/E+/Getty Images
Typesetting: Integra Software Services Pvt. Ltd.

www.degruyter.com

Contents

1 The origins of biotechnology

1.1 Introduction

The term biotechnology is certainly applied very widely today, and yet has grown and expanded to have different meanings in different circles. The United Nations Convention on Biological Diversity defines biotechnology as "any technological application that uses biological systems, living organisms, or derivatives thereof, to make or modify products or processes for specific use [1]." In a similar manner, the Organization of Economic Co-operation and Development (the OECD) has defined it as "the application of scientific and engineering principles to the processing of materials by biological agents [2, 3]."

Very simply put, biotechnology is the use of some living thing, some living organism, to make a material, chemical, or product that is in some way useful to humankind [4]. It now affects the daily life of most people, even though we think of it very little. A common example is what might be called the expanded way in which refuse is disposed of. Some municipalities have separate waste for recyclable materials, and those which can be landfilled. Others go farther, and ask people to segregate their waste by what can be refunded – meaning materials like plastics for which some monetary refund is available at stores and centers. Others ask that materials that can be composted be separated, which often marks such waste for some form of microbial degradation, and possible reuse or further use – a biotechnological solution to this fraction of waste material. Figure 1.1 gives an example.

But as mentioned, this is only one example of a biotechnological solution to a problem, or use of a biological organism to produce or improve some product. Biotechnology finds uses in fields as disparate as fine chemical and pharmaceutical production, leather manufacturing, cosmetics production, and the mining and refining of ores, to name just a few unrelated fields.

1.2 Introduction, historical

Many discussions of biotechnology begin with the two most ancient processes utilized by mankind: brewing beer and baking bread. Both are millennia old; the origins of both are lost in history, and both have been mainstays of human existence almost as long as humanity has existed in settled groups, and not as hunter–gatherers. Both also depend upon organisms that cannot be seen with the naked eye, and that were thus considered somewhat magical throughout much of history.

https://doi.org/10.1515/9783111330259-001

Figure 1.1: Segregated trash receptacles.

1.3 Fermentation

In a very ancient past, fermenting grain appears to have been discovered serendipitously, when stored grain became wet, probably by accident. This appears to have become a process in virtually all ancient civilizations, and the consumption of different types of beer has been both part of everyday eating, as well as part of sacred ritual, throughout much of that time.

Much like the production of beer, the production of wine appears to have occurred in the ancient past essentially by accident. In the case of wine, a species of yeast that lives on the skin of grapes, *Saccharomyces cerevisiae*, promotes the fermentation of grapes. Like beer consumption, the drinking of wine was both for everyday survival (since water sources were often contaminated with some disease-causing organism), as well as for ritual purposes. Concerning the latter, the mixing of water and wine is still part of virtually every Christian denomination's Mass or service.

The fermentation of beer and wine both depends on the presence of *S. cerevisiae* and other yeasts, as well as water and a feedstock, traditionally grain or grapes. Different cultures have used different recipes for the production of beer and wine, and in some cases have incorporated laws about its production. Arguably the most famous case of this in the western world is the German Reinheitsgebot of 1516 (the purity order), which specifies what is allowed as ingredients for the production of beer [5]. It is noteworthy that yeast is not mentioned, because the microorganism and how it affected fermenting grain had not been discovered at the time.

The production of beer and wine is virtually universal, with a large number of international, national, and regional organizations devoted to their development and promotion to the general public [6–20]. Additionally, every major producer of beer, wine, and now other alcoholic beverages have corporate websites advertising their products, usually emphasizing their quality and taste. Because consumer tastes vary widely, and because commercial competition is so keen, grocery store aisles and liquor store aisles now generally offer a very wide selection of beers and wines. Figure 1.2 shows but one example in what can sometimes be a bewildering array.

Figure 1.2: Beer choices.

1.4 Leavening

Baking bread may not be as old as the production of beer; then again, it might be. Ancient records exist for the production of both bread and beer, and indeed, at least one ancient civilization must have known the difference between leavened and un- leavened bread, as is evidenced by the Book of Exodus in the Bible.

Much like the production of beer and wine, the production of bread depends on leavening it, which means the use of yeast to make the bread rise. Bread can certainly be made without yeast, but this results in hard, unleavened breads, often called quick breads. There are a wide variety of unleavened breads, and probably an even larger variety of leavened ones. Leavening, sometimes called the use of a raising agent, can be effected by several different types of yeast, but once again *Saccharomyces cerevisiae* – generally called Baker's yeast in this application – is one of the most common. It produces carbon dioxide, which causes the flour–water mixture of dough to expand.

Since a large portion of the world's population consumes some form of bread each day, it has definitely become one of the leading types of foods produced on an industrial scale. Much like beer and wine production, bread production is so large that there are many national and regional associations that promote its safe, indus- trial-scale production as well as its sale [21–30]. But also, there has been a large growth in small bread manufactures in several countries. The consumer desire for breads made with no preservatives, those that have an artisanal look and feel to them, has become a driving force in the creation of such companies. Figure 1.3 shows a variety of breads marketed at a smaller shop.

Figure 1.3: Artisanal breads.

1.5 Cheese making

A third ancient process that can easily be classified as an early form of biotechnology is the production of cheese. Much like fermenting grain and leavening bread, the origins of cheese production have been lost to time. Evidence of early cheese making has been found in sites as widely separated as Eastern Europe and Central Asia. Unlike today, where a large portion of the cheese produced in the western world is made from cows' milk, sheep milk cheese and goat milk cheese appear to have been the dominant forms in ancient times.

The sources of rennet, the enzyme that ferments and begins the separation of curds from whey, are many, and it is believed that in ancient times it was the act of storing milk in bladders such as animal stomachs that exposed milk to rennet – at first accidentally. There are other theories as to the origin of cheese making as well.

Production of cheese has until recently been a means by which a food can be preserved for long periods of time. As proof of this, more than once a cheese that is more than 100 years old has been found after having been lost in some storage container or facility, and found to be edible. The popular press has even reported on cheeses found in tombs that are over 3,000 years old [31, 32].

Cheese production today has become a large-scale industry, and uses several different types of rennet and molds to give various cheeses their flavors [33]. Rennet is a complex of different enzymes, which includes chymosin and pepsin among others.

It is obtained from the stomachs of animals that must digest their mother's milk. But rennet can now be stored for relatively long periods of time, and thus transported. This becomes important for artisanal cheesemakers who work on a smaller scale than most industries. Figure 1.4 shows a tiny sampling of the many types of cheese that can be produced on a large scale.

Figure 1.4: Cheese varieties.

Cheese making though has become a widespread and large enough industry that there are numerous national and regional organizations dedicated to it, as there are for beer, wine, and bread production [34–42].

1.6 Modern biotechnology and recombinant DNA

The three processes just discussed represent the ancient roots of biotechnology. To them we can add the slow manipulation of plants and animals by selective breeding to produce species with desirable traits over the course of generations. But what most people consider biotechnology as modern has its origin in the early 1970s, when the idea of directly manipulating DNA by making selective changes in it in a laboratory setting was first developed, the first direct bioengineering. Some of the earliest work in this area was performed at Stanford, and the first patent in this area was granted to Professors Stanley Cohen and Herbert Boyer in 1980 – the fathers of the field – although it had been applied for in 1973. This was for a DNA sequence they had produced, and its use in *E. coli* bacteria.

Dr. Boyer was one of the founders of the firm Genentech, which used this technique to produce human insulin as one of its early products. The value of this and other products was recognized relatively quickly, and in 1990 Genentech was essentially acquired by F. Hoffmann-La Roche.

Because this earliest work in the manipulation of DNA was directed at what can broadly be called drug design and the medical field, much of the general public today considers biotechnology an area dominated by and confined to medicine. We will see in subsequent chapters that while there is an enormous amount of biotechnological work in the medical field, there is also a large amount of industrial-level biotechnology used in fields that are far removed from this area. Yet it is indisputable that the pioneering work of Cohen and Boyer has dramatically and forever changed the way industry now produces many materials and products.

References

[1] United Nations Convention of Biological Diversity. Website. (Accessed 7 January 2024, as: https://www.cbd.int/convention/text/).

[2] Organisation for Economic Co-operation and Development. Website. (Accessed 7 January 2024, as: http://www.oecd.org/).

[3] The Application of Biotechnology to Industrial Sustainability. Website, and download. (Accessed 7 January 2024, as: web-archive.oecd.org/2012-06-15/155236-1947629.pdf).

[4] Amgen. Website. (Accessed 7 January 2024, as: https://biotechnology.amgen.com/biotechnology-science.html).

[5] Reinheitsgebot. Website. (Accessed 7 January 2024, as: http://www.brewery.org/library/Rein Heit.html).

[6] The Brewers of Europe. Website. (Accessed 7 January 2024, as: https://www.brewersofeurope.eu).

[7] Brewers Association. Website. (Accessed 9 January 2024, as: https://www.brewersassociation.org/).

[8] Wine America. Website. (Accessed 9 January 2024, as: https://wineamerica.org/).

[9] California Association of Wine Grape Growers. Website. (Accessed 9 January 2024, as: https://www.cawg.org/).

[10] Wine Institute. Website. (Accessed 9 January 2024, as: https://www.wineinstitute.org/).

[11] Michigan Wine Producers Association. Website. (Accessed 9 January 2024, as: http://www.miwpa.org/).

[12] Wine Growers Canada. Website. (Accessed 9 January 2024, as: https://winegrowerscanada.ca).

[13] Grape Growers of Ontario. Website. (Accessed 9 January 2024, as: https://www.grapegrowersofontario.com/).

[14] Wines of Canada. Website. (Accessed 9 January 2024, as: https://winesofcanada.com/).

[15] GermanWine.de. Website. (Accessed 9 January 2024, as: https://www.germanwine.de/english/).

[16] European Wine Growers Association. Website. (Accessed 23 January 2024, as: https://www.ewga.net/).

[17] South Australian Wine Industry Association Incorporated. Website. (Accessed 23 January 2024, as: https://www.winesa.asn.au/).

[18] Wine Grape Growers Australia. Website. (Accessed 23 January 2024, as: https://www.agw.org.au).

[19] New Zealand Wine. Website. (Accessed 23 January 2024, as: https://www.nzwine.com/en).

[20] Wines of South Africa. Website. (Accessed 23 January 2024, as: https://www.wosa.co.za/home/).

[21] U.S. Wheat Associates. Website. (Accessed 23 January 2024, as: https://www.uswheat.org/).

[22] Bread Bakers Guild of America. Website. (Accessed 23 January 2024, as: https://www.bbga.org/).

[23] National Association of Wheat Growers. Website. (Accessed 23 January 2024, as: https://www.wheatworld.org/).

[24] Western Canadian Wheat Growers. Website. (Accessed 23 January 2024, as: https://www.wheatgrowers.ca/).

[25] Michigan Bread. Website. (Accessed 23 January 2024, as: https://michiganbread.com/).

[26] Grain Growers. Website. (Accessed 23 January 2024, as: https://www.graingrowers.com.au/).

[27] Grain S.A. Website. (Accessed 23 January 2024, as: https://www.grainsa.co.za/).

[28] Wheat Growers. Website. (Accessed 23 January 2024, as: https://www.wheatgrowers.com/).

[29] Washington Association of Wheat Growers. Website. (Accessed 23 January 2024, as: https://www.wawg.org/).

[30] Artisan Baker Association. Website. (Accessed 23 January 2024, as: https://sourdough.com).

[31] USA Today. Website. (Accessed 23 January 2024, as: https://www.usatoday.com/story/tech/2014/02/25/worlds-oldest-cheese/5776373/).

[32] International Dairy Foods Association. History of cheese. Website. (Accessed 23 January 2024, as: https://www.idfa.org/resource-center/industry-facts/cheese/).

[33] Jenkins, S. Cheese Primer, Workman Publishing, New York, NY, 1996, 0-89480-762-5.

[34] International Dairy Foods Association. Website. (Accessed 23 January 2024, as: https://www.idfa.org).

[35] Wisconsin Cheese Makers Association. Website. (Accessed 23 January 2024, as: https://www.wischeesemakersassn.org/).

[36] American Cheese Society. Website. (Accessed 23 January 2024, as: https://www.cheesesociety.org/about-us/missionandvalues/).

[37] New York State Cheese Manufacturers Association. Website. (Accessed 23 January 2024, as: https://nyscheesemakers.com/).

[38] Dairy Farmers of Canada. Website. (Accessed 23 January 2024, as: https://dairyfarmersofcanada.ca).

[39] Sustainable Farming Association, Artisanal Cheese Making. Website. (Accessed 23 January 2024, as: http://www.sfa-mn.org).

[40] Dairy Industry Association of Australia. Website. (Accessed 23 January 2024, as: https://diaa.asn.au/).

[41] New Zealand Specialist Cheese Makers Association. Website. (Accessed 23 January 2024, as: https://www.nzsca.org.nz/).

[42] South African Cheese. Website. (Accessed 23 January 2024, as: https://www.cheesesa.co.za).

2 Biotechnological processes today

2.1 Introduction

To say that biotechnology has advanced a long way from the traditional and ancient processes of fermentation to where it is today is a truly enormous understatement. The field has moved from a set of efforts that are best described as trial and error, to one in which specific chemical and genetic information can be transferred to produce desired and specific results. Biotechnological processes have been incorporated into numerous areas within industry, and continue to expand in scope today, as the industry continues to expand at an incredibly fast rate. Organizations and associations of biotechnology-based companies, as well as academic and government stakeholders have been formed as a means to encourage growth in the industry, as well as to advocate for it to regional and national governments [1–14]. Additionally, the comingling of biotechnology with the green chemistry movement has manifested itself in numerous consumer products, and even in kiosks and stores devoted to items connected to the idea. Figure 2.1 gives an example.

Those interests and companies dedicated to the promotion of biotechnology at the industrial scale might point out that such stores and commercial ventures like those in Figure 2.1 appear to be a fad that simply caters to the general public's wish to buy products which appear to have some connection to anything that appears to be environmentally friendly or biologically based. But the counterpoint can be made that the presence of such helps make the public aware of the possibilities of green chemistry and biotechnology.

2.2 Biotechnology companies

The term "biotechnology company" can be somewhat misleading, since some companies exist solely because of their biotechnology efforts at producing one or a select few products, while others are long-established companies that have mature product offerings which existed before the advent of modern biotechnological methods, and that have expanded into the biotech field. In the latter case, some industries have simply added biotech processes to produce new chemicals and materials, while others have incorporated some technique that utilizes one or more biotech steps, simply because it is an economically more profitable process than anything that came before.

The top 25 biotechnology companies is a list that changes annually, as companies expand, merge, or in other ways change. Such a list includes some very large, well-established names, as well as some that have only recently come into existence. Table 2.1 lists these companies, as well as their major products or field of endeavor [15–39].

https://doi.org/10.1515/9783111330259-002

Figure 2.1: Kiosk promoting green products.

Undoubtedly, this list will continue to grow and change, as new companies come into existence, and as companies continue to merge, grow, and split off new corporate entities.

2.3 Biotechnology sectors

While Table 2.1 shows a heavy concentration of biotechnology to companies that provide medicines, pharmaceuticals, and other health-related products, the field of biotechnology can be divided in numerous different ways. In this book we have chosen to divide the subject into the fields of energy, food, and products and materials. Undoubtedly, there are other ways to group these processes – with products and materials being an area that continues to not only grow but to broaden. Only recently have different corporations moved into the use of microorganisms to aid in the early steps of metal refining, for example.

Curiously, several terms have recently been brought into the lexicon that divide biotechnology by what are essentially colors. They are listed in Table 2.2, but are certainly not the only way biotechnological sectors can be defined.

But even with such delineators, as a field, biotechnology has what might be called fuzzy edges – meaning unclear lines as to where it stops. In the wake of the advances of the past 30 years, do incremental advances in fermentation for alcohol production still qualify as biotechnological advances? Does the use of organisms to consume or isolate sulfur from sulfide ores qualify as a biotechnological technique, since the refining of metals has not been a research thrust that has traditionally been part of any biotech firm? It is difficult to give clear answers upon which all active and interested parties will agree. There are some areas though upon which there is agreement. They include the following:

Table 2.1: Biotechnology companies.

Company name	Major products	Comments	Web address
Johnson & Johnson	Medical, pharmaceutical, consumer health care		https://www.jnj.com/
Roche	Hemophilia, multiple sclerosis medicines		http://www.roche.com/
Pfizer	Wide listing of prescription and over-the-counter medications	http://www.pfizer.com/products/drug-development/process/logic-of-a-biologic	http://www.pfizer.com/
Novartis	". . .innovative medicines, eye care, and generics [18]."		https://www.novartis.com/
Merck & Co.	Vaccines, oncology products, prescription medicines		http://www.merck.com/index.html
Amgen	Medical, health care		http://www.amgen.com/
Sanofi	Pharmaceuticals, vaccines		http://www.sanofi.us/l/us/en/index.jsp
AbbVie	Pharmaceuticals	Known for Humira	https://www.abbvie.com/
GlaxoSmithKline	Medications, vaccines, other health-care products		https://www.gsk.com/
Celgene	Pharmaceuticals		http://www.celgene.com/about/
Bayer	Pharmaceuticals		https://www.bayer.com
Eli Lilly & Co.	Pharmaceuticals	Known for Humalog ®	https://www.lilly.com/
Gilead Sciences	Pharmaceuticals, emphasis on oncology		http://www.gilead.com/
Bristol-Meyers Squibb	Pharmaceuticals	Known for Plavix ®	https://www.bms.com/
Allergan	Pharmaceuticals		https://www.allergan.com/home
AstraZeneca	Biopharmaceuticals	Focus on oncology	https://www.astrazeneca.com/

Table 2.1 (continued)

Company name	Major products	Comments	Web address
Abbott Laboratories	Diabetes medicines, nutrition products		http://www.abbott.com/
Novo Nordisk	Diabetes medicines, growth hormone therapies		http://www.novonordisk.com/
Biogen	Multiple sclerosis medicines		https://www.biogen.com/
Shire Pharmaceuticals	Vyvanse, Adderall XR	Slogan: "Championing the fight against rare diseases [34]."	https://www.shire.com/
Stryker Corporation	Reconstructive biosurgery		http://www.stryker.com/en-us/index.htm
Regeneron Pharmaceuticals, Inc.	Dupixent, Praluent	Significant use of modified mice	https://regeneron.com/
Teva Pharmaceutical	Central nervous system disorder medicines, generics, OTCs		http://www.tevapharm.com/
Baxter International	Anesthesia & Critical Care; BioSurgery; Drug Delivery; Nutrition; Renal Therapies [38]		http://www.baxter.com/index.page
Alexion Pharmaceuticals	Soliris, Strensiq, Kanuma		http://www.alexion.com/

Table 2.2: Biotechnology sectors.

Title	Area of specialization
Blue biotechnology	Use of aquatic resources, algae-based oils
Brown biotechnology	Desert development and management, plant production for arid climates
Gray biotechnology	Environmental management and remediation
Green biotechnology	Agriculture
Red biotechnology	Pharmaceuticals and medicine
White biotechnology	Industrial chemical production
Yellow biotechnology	Production of foods

2.3.1 Fuel and energy

Although the production of alcoholic beverages has an ancient past, the use of biomass to produce what is now called bioethanol and biodiesel for automotive fuels, as well as oils for other fuel uses, is a relatively new use for this large-scale fermentation. Table 2.3 lists several of the leading companies in this area, but by no means all such companies. The use of corn, sugarcane, soybeans, algae, and other plant and animal sources for fuel is an area that has seen enormous growth in the past 30 years.

Table 2.3: Biotechnology fuel companies.

Company name	Major products	Comments	Web address
AltAir Fuels	Jet fuel		www.altairfuels.com
Dynamic Fuels	Diesel	Animal fat and vegetable oil sourced	www.dynamicfuelsllc.com
Envergent Technologies	Jet fuel		www.envergenttech.com
Gevo	Isobutanol		www.gevo.com
LanzaTech	Biocommodities	Aims to produce fuels without using food sources	www.lanzatech.co.nz
Neste Oil	Renewable diesel		www.nesteoil.com
Oilgae	Diesel	Algae sourced	www.oilgae.com
SG Biofuels	Plant oils	Fuels from japtropha seed	www.sgfuel.com
Sapphire Energy	Gasoline, diesel, jet fuel	Algae sourced	www.sapphireenergy.com
Solazyme	Renewable fuels	Algae sourced	www.solazyme.com
Terrabon	Gasoline	Nonfood sourced	Terrabon.com

2.3.2 Food and food supplements

A wide variety of foods have been genetically altered in the past few decades, to impart some trait in them that was not present before. Certain plants have been adjusted and enhanced for millennia, with techniques that predate any biotechnological methods. The production of corn is an excellent example. What is considered normal corn-on-the-cob today is actually a relatively recent plant, one that has been made possible largely by seed selection over the course of decades – indeed, over the course of centuries. But plants that are genetically modified, such as the now famous "Flavr Savr" tomato, have been deliberately modified to enhance specific traits. This is the result of modern biotechnological techniques, and not through trial and error.

Additionally, the twentieth century has seen an explosion of what are called food additives. Many foods have preservatives added to them, but the entire field of vitamin production is one that has arisen in the relatively recent past. Diseases such as scurvy have been eradicated because of the ability to produce vitamin C on a large scale, as well as the ability to distribute it to people throughout the world. But the synthesis of vitamin C – as opposed to its extraction from citrus fruits – requires the use of *Gluconobacter oxydans* (*G. oxydans*), sometimes called *Acetobacter*, in one step to ensure the final product is stereochemically correct. Similarly, the much more structurally complex vitamin B_{12} requires fermentation of either *Pseudomonas denitrificans* or *Propionibacterium freudenreichii* for its large-scale synthesis.

2.3.3 Drugs and other pharmaceuticals

As shown in Table 2.1, drugs and pharmaceuticals are the apparent main focus of many biotech companies. The reason for this is straightforward: as the quality of life and human health has increased, more complex drugs and medications are required, both for treatment and for diagnosis. This in turn means that biotechnological means of production are often far more effective than what might be called traditional organic synthesis. Perhaps the single most poignant example of this is the just-mentioned production of vitamin B_{12}. The most complex of the vitamins, the large-scale production of vitamin B_{12} currently cannot occur without specific fermentation of microorganisms. Interestingly, Professor R.B. Woodward did dedicate the manpower and time to a total synthesis of vitamin B_{12}, one that reportedly took more than 100 researchers to complete [40]. Clearly, the biotech route is a more efficient and more profitable one.

The use of biotechnological methods in the large-scale production of drugs has seen rapid expansion in the recent past, but also has a relatively long history. If one considers the production of medicinal alcohol the production of a drug, this history stretches back to the dawn of human-induced chemical transformation of raw materials. But it is fair to claim that the history of what has become large-scale pharmacology has its origins in the early nineteenth century.

2.3.4 Industrial products

A large number of products that the general public does not normally associate with biotechnology have been developed and brought to large-scale production quite recently. These include chemicals used in the production of several wood products, as well as chemicals used for the production of leather goods. Different types of rubber also now utilize biotechnological techniques in their production. Even some metal ores begin their transformation to reduced metals through the use of microorganisms reacting with their ores to capture such species as sulfides, species that would otherwise be oxidized to sulfur oxides (airborne pollutants and a persistent problem).

References

[1] European Biotechnology Network. Website. (Accessed 23 January 2024, as: https://european-biotechnology.net).

[2] European Association of Pharma Biotechnology. Website. (Accessed 23 January 2024, as: https://www.bionity.com).

[3] EuropaBio. Website. (Accessed 23 January 2024, as: https://www.europabio.org/).

[4] EuropaBio: The European Association of Bioindustries. Website. (Accessed 23 January 2024, as: https://www.europabio.org/library/).

[5] Japan Bioindustry Association. Website. (Accessed 23 January 2024, as: https://www.jba.or.jp).

[6] Young European Biotech Network. Website. (Accessed 23 January 2024, as: http://www.yebn.eu/).

[7] BIO. Biotechnology Innovation Association. Website. (Accessed 23 January 2024, as: https://www.bio.org).

[8] BIOTECanada. Website. (Accessed 23 January 2024, as: https://www.biotech.ca/).

[9] Swiss Biotech Association. Website. (Accessed 23 January 2024, as: https://www.swissbiotech.org).

[10] efpia. European Federation of Pharmaceutical Industries and Associations. Website. (Accessed 23 January 2024, as: http://www.ebe-biopharma.eu/).

[11] New England Biotech Association. Website. (Accessed 23 January 2024, as: http://www.newenglandbiotech.org/).

[12] Illinois Biotechnology Innovation Organization (iBIO). Website. (Accessed 23 January 2024, as: https://www.bio.org/).

[13] Iowa Biotech Association. Website. (Accessed 7 February 2024, as: https://www.iowabio.org/).

[14] New Mexico Biotechnology & Biomedical Association. Website. (Accessed 7 February 2024, as: https://www.nmbio.org/).

[15] Johnson & Johnson. Website. (Accessed 7 February 2024, as: https://www.jnj.com/).

[16] Roche. Website. (Accessed 7 February 2024, as: https://www.roche.com/).

[17] Pfizer. Website. (Accessed 7 February 2024, as: https://www.pfizer.com/).

[18] Novartis. Website. (Accessed 7 February 2024, as: https://www.novartis.com).

[19] Merck & Co. Website. (Accessed 7 February 2024, as: https://www.merck.com).

[20] Amgen. Website. (Accessed 7 February 2024, as: https://www.amgen.com/).

[21] Sanofi. Website. (Accessed 7 February 2024, as: https://www.sanofi.us/).

[22] AbbVie. Website. (Accessed 7 February 2024, as: https://www.abbvie.com/).

[23] GlaxoSmithKline. Website. (Accessed 7 February 2024, as: https://www.gsk.com/).

[24] Celgene. Acquired by Bristol Myers Squibb in 2022 – accessed 7 February 2024.

[25] Bayer. Website. (Accessed 7 February 2024, as: https://www.bayer.com/).

[26] Eli Lilly & Co. Website. (Accessed 7 February 2024, as: https://www.lilly.com/).

[27] Gilead Sciences. Website. (Accessed 7 February 2024, as: https://www.gilead.com/).

[28] Bristol-Meyers Squibb. Website. (Accessed 7 February 2024, as: https://www.bms.com/).

[29] Allergan, part of AbbeVie. Website. (Accessed 7 February 2024, as: https://www.abbvvie.com/allergan.html).

[30] AstraZeneca. Website. (Accessed 7 February 2024, as: https://www.astrazeneca.com/).

[31] Abbott Laboratories. Website. (Accessed 7 February 2024, as: https://www.abbott.com/).

[32] Novo Nordisk. Website. (Accessed 7 February 2024, as: https://www.novonordisk.com/).

[33] Biogen. Website. (Accessed 7 February 2024, as: https://www.biogen.com/en_us/home.html).
[34] Shire Pharmaceuticals – acquired by Takeda Pharmaceutical Company, 8 January 2019. Website. (Accessed 7 February 2024, as: https://www.takeda.com/).
[35] Stryker Corporation. Website. (Accessed 7 February 2024, as: https://www.stryker.com).
[36] Regeneron Pharmaceuticals, Inc. Website. (Accessed 7 February 2024, as: https://www.regeneron.com/).
[37] Teva Pharmaceutical. Website. (Accessed 7 February 2024, as: https://www.tevapharm.com/).
[38] Baxter International. Website. (Accessed 7 February 2024, as: https://www.baxter.com/).
[39] Alexion Pharmaceuticals. Website. (Accessed 7 February 2024, as: https://alexion.com/).
[40] Khan, A.G. and Eswaran, S.V. Woodward's synthesis of vitamin B12. Resonance, 2003, 8(6), 8–16. Website. (Accessed 7 February 2024, as: https://link.springer.com/article/10.1007/BF02837864).

3 Bioethanol

3.1 Introduction

Ethanol has been produced by the fermentation of grains, fruits, and vegetables for millennia, as discussed in Chapter 1. Throughout almost all of history, people were unaware that some form of yeast caused the change from the starches and sugars of a plant-based starting material to the final ethanol. They simply followed recipes that stretched back into a distant past, into antiquity [1].

In the latter half of twentieth century, ethanol production on an industrial scale used two-carbon molecules that had been refined and isolated from crude oil as a feedstock, primarily ethylene. The shift back to large-scale feedstocks such as corn and sugarcane is a relatively recent one, and has grown not only because of demand for ethanol as fuel, but as a means to keep the price of corn steady in some parts of the world. Today it is a large enough operation globally that there are several trade organizations either totally or partially devoted to its production [2–12]. Figure 3.1 shows a basic scheme for how ethanol is produced from ethylene starting material.

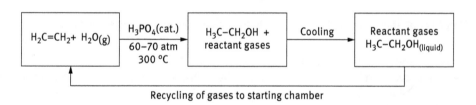

Figure 3.1: Ethanol production from ethylene recycling of gases to starting chamber.

It is shown in Figure 3.1 that despite high temperatures and pressures, and the use of a catalyst, the reaction still does not run to completion in a single pass. Thus, a reaction with a significant capital input requires even more energy to obtain complete conversion. Added to this is the source of ethylene, crude oil, a nonrenewable feedstock. It is not difficult to imagine then that some less energy-intensive process, preferably one using a renewable feedstock, is a preferential way to produce this commodity. The bio-based production of ethanol via fermentation, using some starch-containing or cellulose-containing starting material, is certainly a move that reduces energy input, and removes the need for a fossil fuel source. This has become a matter of interest to governments, some of which have formed committees to monitor how bioethanol is produced [13, 14].

In this chapter, we will discuss the production of fuel ethanol from both starch sources as well as cellulosic sources.

https://doi.org/10.1515/9783111330259-003

3.2 Bioethanol from starch

As mentioned in Chapter 1, the fermentation of grain has been the longest-standing form of ethanol production, with a history going back to several ancient civilizations. Currently, the major production methods for biobased ethyl alcohol start with corn, sugarcane, and sugar beets. The fermentation of these starch sources accounts for virtually all of the fuel ethanol on the market today, with only small contributions coming from other plant sources. Figure 3.2 shows the basic chemistry involved.

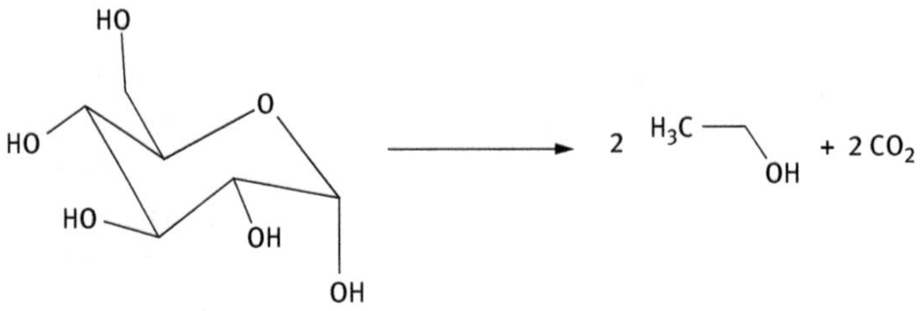

Figure 3.2: Ethanol production from starch.

What is not shown in such a reaction is the yeast that must be present to effect the reaction, and that the reaction must be held at 30–40 °C, at a pH $\approx$ 4. Some yeasts are proprietary to a specific company, although a wide range of yeasts have the enzymes in them to break starch down into ethanol. Additionally, what are now called brewers' yeasts have been used for thousands of years to initiate such fermentations [1]. The chemistry involved in producing beer and wine is the two traditional processes that require starch and such yeasts. Figure 3.3 shows a fermentation room in an artisanal winery, where the final stage of the process occurs.

It is noteworthy that in this reaction, gaseous CO_2 is allowed back into the atmosphere in some cases, because trapping it in an alcohol solution may force overoxidation to acetic acid. Releasing it to the atmosphere is not considered a major problem, since it is part of the carbon cycle for growing a next generation of plants.

3.3 Cellulosic bioethanol

The difference between the linkages that make up starch and that which make up cellulose is quite small, but is crucially important in how each material exists, and in what enzymes degrade them. As mentioned, starch is easily hydrolyzed by humans and by many other animals, yet cellulose is not. Figure 3.4 shows the linkage in cellulose that cannot be digested by humans and other mammals, while Figure 3.5 shows the starch linkage.

Figure 3.3: Wine fermentation.

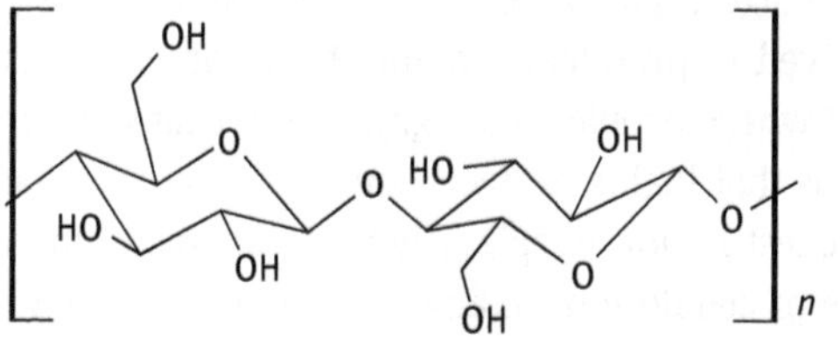

Figure 3.4: β-1,4 Linkage in cellulose (D-polyglucose).

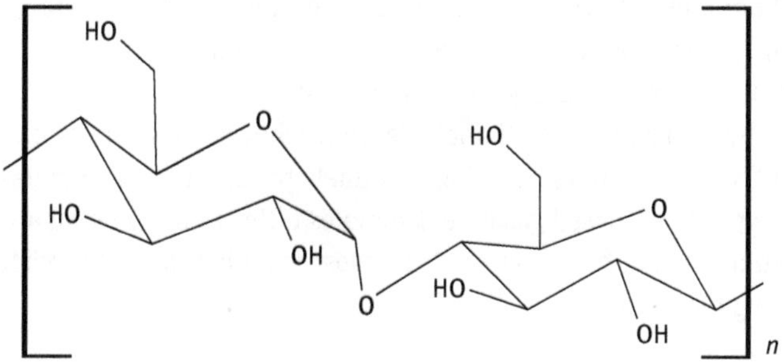

Figure 3.5: α-Glucose linkage in starch.

Several companies have begun some research effort aimed at scaling up the production of ethanol from cellulose to an industrial-scale process. Virtually all use enzymes or yeasts that are proprietary, simply because of the perceived amount of profit that can be made through a successful process. One process that is fairly well known, however, involves termites, or at least their gut bacteria.

Curiously, although termites ingest wood, and thus cellulose, their digestion of it is a symbiotic process involving gut protozoa, which in turn have bacteria on them, which ultimately can produce enzymes that are able to break down cellulose. These are not present in a termite's digestive system at birth, and therefore a termite, slightly after hatching, consumes the excrement of another termite one time. This allows the introduction of the symbiotic organisms into the termite's digestive system and enables it to continue to utilize cellulose in its diet. These gut organisms, such as *Trichonympha* are then in the termite, and able to effect the breakdown of cellulose [15].

With this knowledge, it is perhaps a short, logical step to establishing reaction vessels that contain the correct microbial organism, to determine if ethanol can be produced from cellulose directly. As mentioned, firms that have chosen to explore this currently all keep the details of their operation proprietary, largely because of the potential profit to be made by the company that is successful in such an endeavor.

Any ethical debate about using a food crop for fuel production is avoided when ethanol is produced from the cellulosic portion of corn or any other crop. The reason for this is perhaps obvious: humans cannot eat and digest cellulose or cellulose-based materials.

The issue of the ethical debate involved in producing ethanol from food crops is not a trivial one, since we live in a world where people go hungry, and during wars or famines sometimes starve. It seems callous that fuel is "grown" from a food crop, to be used as a motor fuel for perhaps the richest billion people, when the poorest billion may be hungry or starving. Thus, the use of cellulose from any plant source is a move in a more humane direction, since it not only uses a plant source for ethanol, but also frees up more food crops for human consumption.

The use of bioethanol for automotive fuels has nevertheless increased in the past decade, first with 5% ethanol added to regular motor fuel, and then with 10% added. Figure 3.6 shows an example of such a notification at a service station.

Most recently, one can find progressively more service stations offering what is called E-85 fuel, automotive fuels that are 85% ethanol. Figure 3.4 shows one such station.

Note in Figure 3.7 that the station offers both E-85, routinely produced via fermentation, as well as traditional, petroleum-based gasoline. Estimates differ as to when all automobiles will be converted to ethanol fuels, but even the most optimistic do not predict a large-scale switch occurring before 2050.

3.4 Future challenges

Perhaps the greatest challenge to expanding the production of ethanol for use as fuel is not biotechnological, but rather is in finding arable land that is not being otherwise

Figure 3.6: Notice of 10% ethanol in gasoline.

Figure 3.7: Gas station that offers ethanol fuel.

used, and finding both the water and fertilizer to produce enough of any starchy plant, such as corn or sugar cane. The continental United States has essentially all of its arable land already used in some way for the production of food, either for humans or for animals that will eventually become food for humanity. Likewise, the fields and arable land of Europe, of Central Asia, and of Australia are essentially all in use.

Additionally, and very importantly, increasing the scale of bioethanol production from cellulosic sources remains a major challenge. Currently, most woody material that can be considered scrap is already used, albeit not always in the most efficient manner. For example, bark is used, but in many cases simply as a source of fuel for some process that requires heat. As well, small-sized scraps of wood are used to make low-grade particle board. The use of any such material for the production of bioethanol (or for any other higher value product) can be an effective use.

References

[1]	Beer in the Ancient World. Website. (Accessed 22 May 2017, as: http://www.ancient.eu/article/223/).
[2]	Renewable Fuels Association. Website. (Accessed 7 February 2024, as: http://www.ethanolrfa.org/).
[3]	Ohio Ethanol Producers Association. Website. (Accessed 8 February 2024, as: http://ohethanol.com).
[4]	Wisconsin Biofuels Association. Website. (Accessed 11 February 2024, as: http://wibiofuels.org/).
[5]	European Biomass Industry Association. Website. (Accessed 11 February 2024, as: https://www.eubia.org/).
[6]	Advanced Biofuels Association. Website. (Accessed 11 February 2024, as: http://advancedbiofuelsassociation.com/).
[7]	Adavanced BioFuels USA: Renewable Fuel Information A-Z. Website. (Accessed 11 February 2024, as: https://www.advancedbiofuelsusa.info).
[8]	Association of the German Biofuels Producers, VDB. Website. (Accessed 11 February 2024, as: https://www.euractiv.com/content_providers/association-of-german-biofuels-producers-vdb).
[9]	Southern African Bioenergy Association (SABA). Website. (Accessed 11 February 2024, as: biofpr.com/details/link/100232/Southern_African_Biofuels_Association.html).
[10]	Biofuels Association of Australia. Website. (Accessed 11 February 2024, as: https://biofuelsassociation.com.au/).
[11]	Biomass Magazine. Website. (Accessed 11 February 2024, as: https://www.biomassmagazine.com/).
[12]	Fuel Ethanol Workshop & Expo. Website. (Accessed 11 February 2024, as: https://few.bbiconferences.com).
[13]	U.S. Senate, Committee on Energy and Natural Resources. Website. (Accessed 11 February 2024, as: https://www.energy.senate.gov/public/).
[14]	Brazil: Biofuels Annual – USDA Foreign Agricultural Service. Website. (Accessed 11 February 2024, as: https://fas.usda.gov/data/brazil-biofuels-annual-10#).
[15]	Science Daily. Termite Guts May Yield Novel Enzymes For Better Biofuel Production. Website. (Accessed 11 February 2024, as: https://www.sciencedaily.com/releases/2007/11/071121145002.htm).

4 Biodiesel

4.1 Introduction

Diesel engines have a long history, going back into the nineteenth century. They represent something of an odd situation that has not occurred many times in history, namely, that an object was produced – an engine, in this case – that already had a fuel upon which it would function. Admittedly though, the same can be said for gasoline-powered engines. In both cases, the fuel existed, and was being used in some other application when the engine was first produced. Broadly, what makes a diesel engine different from other motor engines is that it can function without the use of a spark plug.

4.2 Petrodiesel production

The traditional refining of crude oil yields a fraction that is used as diesel fuel. This fraction of crude oil distillation is generally 8–21 carbon atoms long. Roughly three-fourths of the mix is a blend of saturated hydrocarbons, with the remainder being some mixture of aromatic hydrocarbons.

Traditionally, diesel fuel's ability to burn has been characterized by what is called a cetane number. The higher the number, the more easily the fuel combusts when sprayed into air that has been heated and compressed. Interestingly, while the bulk of the discussion in this chapter deals with biodiesel, an evolution from what is now called petrodiesel, petrodiesel is itself an evolution from what came before – whale oil. The word "cetane" comes from the Latin word for whale. Prior to the large-scale use of petrodiesel, a great deal of liquid fuel produced for heat and light in the nineteenth century was whale oil. And while we now think of petrodiesel and other petro-fuels in terms of their pollution and their use of a nonrenewable commodity – crude oil – had industry not shifted to petroleum from whale oil in the mid- to late nineteenth century, estimates are that the Earth's oceans would be completely devoid of any whale species today. In short, they would have been hunted to extinction.

The production of diesel fuel today is on par with that of the production of gasoline, with regular gasoline being considered more the fuel of choice for passenger cars and diesel the choice for larger vehicles, such as trucks. The sheer size of such production means it is not difficult to believe that numerous national and international organizations exist, which are dedicated to the promotion and production of the fuel. Perhaps the most obvious is OPEC [1]. But there are several others as well, throughout North America, Europe, Asia, and Australia [2–10].

https://doi.org/10.1515/9783111330259-004

The world dependence upon fuels such as diesel means that there are significant efforts underway to attempt to find renewable sources and means of producing it. Additionally, governments monitor and track the production of diesel as well as gasoline, and have a strong interest in the development of renewable sources, which often means sources that will require a biotechnological step or steps in their production [11–14]. As an example, the United States Senate Committee on Energy and Natural Resources' Subcommittee on Energy states at its website that its jurisdiction includes:

> oversight and legislative responsibilities for: nuclear, coal and synthetic fuels research and development; nuclear and non-nuclear energy commercialization projects; nuclear fuel cycle policy; DOE National Laboratories; global climate change; new technologies research and development; nuclear facilities siting and insurance program; commercialization of new technologies including, solar energy systems; Federal energy conservation programs; energy information; liquefied natural gas projects; oil and natural gas regulation; refinery policy; coal conversion; utility policy; strategic petroleum reserves; regulation of Trans-Alaska Pipeline System and other oil and gas pipeline transportation systems within Alaska Arctic research and energy development; and oil, gas and coal production and distribution. [14]

While such a statement and charge is broad and encompassing, it does indicate how wide the governmental interest is in both the production and use of diesel and other fuels.

4.3 Biodiesel

As with bioethanol, numerous vegetable and some animal species have been used as feedstocks for the production of biodiesel. A partial list of major companies that have invested in biodiesel in the past decade includes the following:
1. ConocoPhillips
2. Neste Oil
3. Valero
4. Dynamic Fuels
5. Honeywell UOP
6. Preem – Evolution Diesel

In all the above-listed companies, their techniques for large-scale production of biodiesel tend to be proprietary, largely because of the potential economic benefits. However, there is a large enough interest in biodiesel production that several regional, national, or transnational organizations have been formed to advocate for it, and in some cases fund its development [15–25].

Despite the inability to share proprietary, corporate information, it has generally been found and published that base-catalyzed reactions are quicker and more cost efficient than acid-catalyzed reactions. The simplified reaction chemistry for the production of biodiesel can be represented as shown in Figure 4.1.

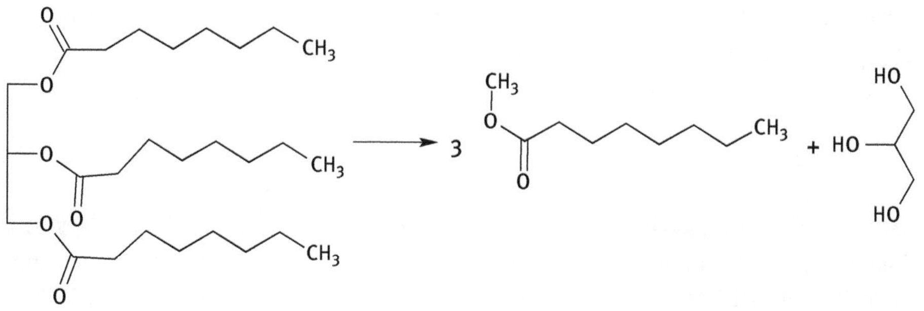

Figure 4.1: Biodiesel production.

An alcohol must be used to effect the transesterification of a fatty acid. Both methanol (shown in the ester in Figure 4.1) and ethanol have found wide use in this regard. Yet this means that to be truly renewable, the alcohol must also have some plant-based or animal-based source. Also note that glycerol, considered the by-product of the reaction, is made in a 1:3 ratio with the resultant fatty methyl esters or ethyl esters. As the production of biodiesel has increased globally in the past two decades, the price of glycerol has been severely depressed within the market for commodity chemicals.

The widespread use of biodiesel has become a point of pride for numerous companies. This allows them to advertise that their fleet vehicles are "greener" than those running on traditional petrodiesel, as does lighter vehicle weight, and sturdy construction from lightweight alloys and composites. An example is shown in Figure 4.2. Of note in

Figure 4.2: Semitractor trailer promoting greener posture.

Figure 4.2 is that the claim of CO_2 reduction can be the result of lighter-weight materials as well as better functioning engines, as well as the use of biodiesel.

4.3.1 Animal fat sources

Animal fats and renderings have found a large variety of uses in the past 70 years, from components of personal care products to blends that are used to feed other animals. But animal fats can easily be transesterified into fatty acids or esters, and glycerol. The example shown in Figure 4.1 is representative of this chemistry, but the length of the fatty acids does not have to be eight carbon atoms long, nor do all three have to be the same length within a single triglyceride. The European Biomass Association estimates that 7% of the biomass consumed is animal fats [17].

The use of animal fats for the production of biodiesel also presents another challenge, this being one of business. Long before there was any interest in using any animal fat source for biodiesel, waste fats were utilized by the cosmetics industry, as well as by the silk screening industries, and in pet food. In cosmetics, animal fats provide a certain feel and adherence to the cosmetic product. In silk screening, it helps the image adhere to a surface. In pet food, it is added as a flavor or flavor enhancer. All of these established industries thus utilize animal fats, and have an economic interest in the material not being used as a source for the production of biodiesel.

4.3.2 Soybean feedstock

Brazil has become a large-scale producer of soybeans, both for food and for biodiesel, because it has a large land mass with areas of low density in terms of human populations, and has soil that is suitable for the growth of soybeans.

The actual material that produces biodiesel when soybeans are the source is soybean oil. The beans can be over 15% oil. The chemistry illustrated in Figure 4.1 is essentially that of the production of biodiesel from soybeans. Figure 4.3 illustrates the steps to arrive at soybean oil. Note that the oil is ultimately extracted with an organic solvent, meaning a continued source of the solvent (hexane) must be available constantly.

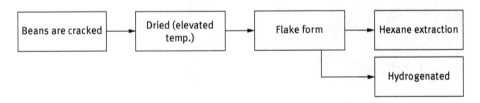

Figure 4.3: Steps in the production of biodiesel from soybeans.

A flow diagram like that in Figure 4.3 must omit the economic aspect of the process, meaning an examination of the costs involved. In brief, 50–75% of the cost of biodiesel is the cost of the vegetable oil, largely because this must be refined and cleaned of water before conversion to fuel. The cost of methanol and/or ethanol also must be factored in, especially as the scale of such an operation is increased. Likewise, the cost of drying a feedstock like soybeans must be factored into the cost of the end product.

4.3.3 Algae

A large number of different types of algae have been tested for their ability ultimately to produce biodiesel. The initial testing of any form of algae dates back to the late 1970s, and an effort by the United States National Renewable Energy Laboratories [26]. Algae are simple to grow, requiring light and water, and do not require land that would compete with food products. The water must be clean enough to grow the plants, and contain the necessary fertilizers. The light can be natural sunlight, or can be provided as fluorescent light tubes immersed in tanks where the algae is being grown. As well, some algae are almost 50% fat and oils.

The general steps by which algae can be grown and turned into biodiesel are shown in Figure 4.4.

To date, although several companies have made serious efforts at large-scale production of biodiesel with algae, the costs still remain too high for it to be economically competitive with petrodiesel. This has not stopped researchers however, as new strategies for its production continue [27].

4.4 Challenges for the future

Much like ethanol, the major challenge in the production of biodiesel is finding enough arable land that can be dedicated to the production of crops for biodiesel, without taking away food sources for humanity. While the use of animal sources for biodiesel has proven itself to be viable, it does not appear that there is enough unused animal by-products from meat processing and packing to serve as a viable large-scale source of fatty materials for biodiesel.

Curiously, at the present time, biodiesel can be made to such a level of purity that when stored for long periods of time, it can actually begin to degrade – to rot. In this state, it is pure enough that some microorganisms can digest it. Thus, it requires the addition of a small amount of some material that prevents this. It has been found that 1% of petrodiesel, when added to biodiesel, prevents such degradation.

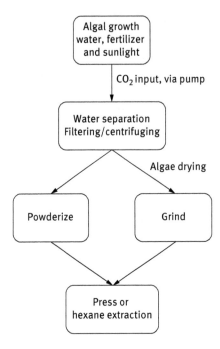

Figure 4.4: Biodiesel from algae.

References

[1] OPEC. Website. (Accessed 11 February 2024, as: https://www.opec.org/opec_web/en/).
[2] IPAA. Independent Petroleum Association of America. Website. (Accessed 11 February 2024, as: https://www.ipaa.org/).
[3] Canadian Association of Petroleum Producers. Website. (Accessed 11 February 2024, as: https://www.capp.ca/).
[4] OEUK. Website. (Accessed 5 March 2024, as: https://oeuk.org.uk).
[5] Fuels Industry UK. Website. (Accessed 5 March 2024, as: http://www.fuelsindustryuk.org).
[6] Fuels Europe. Website. (Accessed 5 March 2024, as: https://www.fuelseurope.eu/).
[7] World Petroleum Council. Website. (Accessed 5 March 2024, as: https://www.wpcenergy.org).
[8] IOGP Europe: International Association of Oil & Gas Producers. Website. (Accessed 5 March 2024, as: https://www.iogpeurope.org).
[9] APPEA. Website. (Accessed 10 October 2018, as: https://www.appea.com.au/about-appea/members/).
[10] Federation of India Petroleum Industry. Website. (Accessed 10 October 2018, as: http://www.fipi.org.in/).
[11] U.S. Senate, Committee on Energy and Natural Resources. Website. (Accessed 12 May 2017, as: https://www.energy.senate.gov/public/).
[12] British Parliament, Commons Select Committee, Energy and Climate Change Committee. Website. (Accessed 10 October 2018, as: https://www.parliament.uk/business/committees/committees-a-z/commons-select/energy-and-climate-change-committee/).
[13] U.S. Senate. Committee on Commerce, Science and Transportation. Website. (Accessed 10 October 2018, as: https://www.commerce.senate.gov/public/).

[14] U.S. Senate. Committee on Energy and Natural Resources. Website. (Accessed 10 October 2018, as: https://www.energy.senate.gov/public/).

[15] Renewable Fuels Association. Website. (Accessed 12 May 2017, as: http://www.ethanolrfa.org/).

[16] Wisconsin Biofuels Association. Website. (Accessed 12 May 2017, as: http://wibiofuels.org/).

[17] AEBIOM, European Biomass Association. Website. (Accessed 12 May 2017, as: http://www.aebiom. org/).

[18] Advanced Biofuels Association. Website. (Accessed 12 May 2017, as: http://advancedbiofuelsassocia tion.com/).

[19] From Feedstock to Fuels. Website. (Accessed 12 May 2017, as: http://advancedbiofuelsassociation. com/page.php?sid=2&id=24).

[20] Association of the German Biofuels Industry, VDB. Website. (Accessed 12 May 2017, as: https://fuel sandlubes.com/tag/association-of-the-german-biofuels-industry-vdb/).

[21] South African Bioenergy Association. Website. (Accessed 12 May 2017, as: http://www.saba.za.org/).

[22] Biofuels Association of Australia. Website. (Accessed 12 May 2017, as: http://biofuelsassociation.com. au/).

[23] Biomass Magazine. Website. (Accessed 12 May 2017, as: http://www.biomassmagazine.com/).

[24] Biodiesel Magazine. Website. (Accessed 10 October 2018, as: http://www.biodieselmagazine.com/).

[25] World Ethanol and Biofuels Reports. Website. (Accessed 12 May 2017, as: https://www.agra-net.com/ agra/world-ethanol-and-biofuels-report/brazil – industrial-biotechassociation-to-be-launched– 1.htm).

[26] National Renewable Energy Laboratory. Website. (Accessed 11 October 2018, as: NREL https://www. nrel.gov).

[27] U-M Team to Create Diesel Fuel from Algae. Detroit Free Press, 11 October 2018, p. B1.

5 Biobutanol

5.1 Introduction

Bioethanol and biodiesel have become the major biofuels for automobiles in the past decades, but other possible biobased fuels exist as well. One of them is biobutanol. This four-carbon member of the broad family of hydrocarbons exists either as a straight-chain molecule or as a branched molecule. There are four possible isomers [1], shown in Figure 5.1, with the third, isobutanol, being the one best suited for use as a motor fuel.

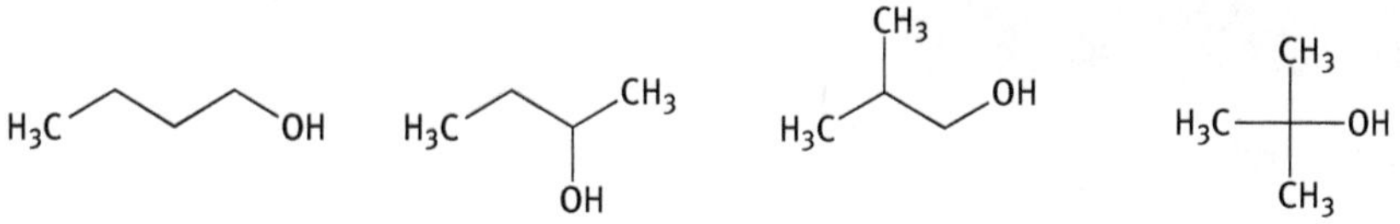

Figure 5.1: Butanol isomers.

Butanol, from petroleum or biobased sources, has been used effectively as a solvent for a wide number of industrial processes. Its relatively high vapor pressure is such that its separation from a product is not particularly energy intensive. Thus, butanol has an already existing set of uses in industry, and the consideration of it as a fuel source, or component of a motor fuel blend, is simply an outgrowth of an existing suite of uses.

Butanol can also be blended into gasoline mixtures, and it is in this role that biobutanol becomes useful as a next generation component of motor gasoline. Up to one-eighth of a gasoline blend can be biobutanol, according to the Clean Air Act [2, 3]. As well, one of the attractions of biobutanol, as opposed to bioethanol, is that engines do not have to be adapted or changed to run on this fuel, or a fuel blend. The term for such compatibility is that butanol is a "drop-in" fuel. Currently, special engine requirements are needed for automobiles that use ethanol of any source, and are usually on these automobiles. Figure 5.2 shows an example.

5.2 Sources

Biobutanol is often made through the same distillation process that produces bioethanol, which means that several plant sources can be used, although corn is used predominantly in the United States simply because so much corn is grown there. However,

https://doi.org/10.1515/9783111330259-005

Figure 5.2: Label indicating an automobile that runs on gasoline and ethanol.

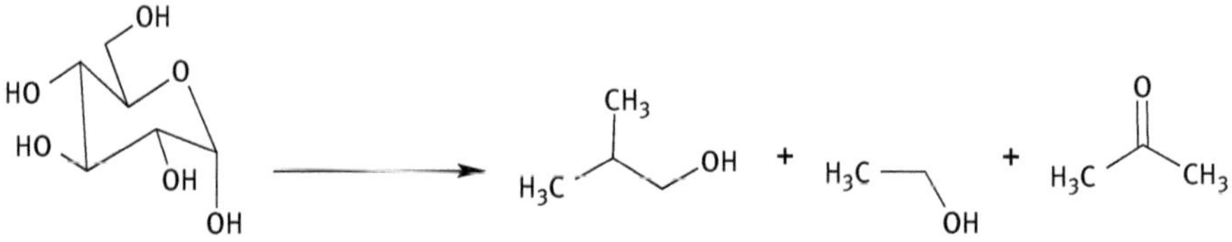

Figure 5.3: Biobutanol production.

beets and other vegetables can be used for its production. Figure 5.3 shows the simplified reaction that starts with glucose and results in butanol.

Note that ethanol and acetone are coproduced in this anaerobic fermentation. Both products must be separated from the butanol product, but both are themselves usable products that add value to any such operation.

5.3 Production

Biobutanol is produced in much the same manner as bioethanol, in what is often called an A.B.E. fermentation (A.B.E. meaning acetone–butanol–ethanol). As mentioned, Figure 5.3 indicates that ethanol is coproduced with butanol, in amounts that are dependent upon the yeast or other organism used. The bacterium most commonly used in the past is some strain of *Clostridia*. While several strains can ferment the starting sugars, *Clostridium acetobutylicum* has been used extensively. This is still sometimes referred to by an older, less scientific term, a "Weizmann Organism," called so after its discoverer [4–7].

A.B.E. fermentations are akin to the fermentation that produces ethanol and carbon dioxide from starch, as mentioned in Section 3.2, but are anaerobic. The general ratio of product is 3:6:1, acetone–butanol–ethanol, but again, can be different based on the yeast and method of production.

Two companies that produce butanol on a large scale, starting from biomass, are Butamax [8] and Gevo [9]. The first is a joint operation between DuPont and BP. The second has as one of its founders Professor Frances Arnold, a recipient of the 2018 Nobel Prize in chemistry. Arnold received the Prize for her work in what is called "directed evolution" of various enzymes. Both companies keep certain information about their butanol production processes proprietary, but a general schematic of such a process is shown in Figure 5.4.

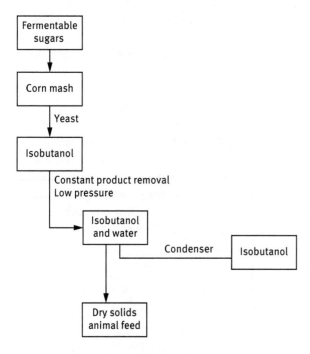

Figure 5.4: Biobutanol production scheme.

The Gevo website does indicate that the continuous removal of product, meaning butanol, from the system is a key to keeping the special yeast alive during the process. As with traditional ethanol production from fermentation, the yeast will eventually die in the product they produce, so removal of it maximizes the yield while keeping the yeast alive. Also note in Figure 5.4 that the solids portion of the final products has a use, usually as animal feed. This then is a second product that generates value from the process.

5.4 Advantages to biobutanol

Since bioethanol is the current biofuel of greatest use, the adoption and use of biobutanol on a large scale will require distinct advantages. These will have to overcome the costs associated with start-ups for biobutanol operations, since bioethanol operations are already mature and in place [8, 10].

- Drop-in fuel: Biobutanol added to current motor fuel mixes does not require any new vehicle engine design
- Energy content: While the energy density of butanol is approximately 15% lower than traditional gasoline, it is higher than fuel additives such as ethanol.
- Sugar sources: Butanol can be produced from a wide variety of different renewable plant feedstocks.
- Minimal retrofitting: Several variations of existing bioethanol plants can be converted to biobutanol without costly retrofitting for the new product.

The first two are two powerful advantages for the use of biobutanol when compared to other alcohols, specifically bioethanol. The third and fourth are largely economic advantages, and so are important as this fuel's production is scaled up and its use broadened. Beyond these, the advantages of any biosourced fuel are roughly the same as any other. For example, all have lower emission than traditional gasoline, because the CO_2 produced is simply that which was captured for the growth of plant material at the beginning. Likewise, the economic advantage of creating jobs domestically (which this scale up will do), for any country, is the same whether butanol, ethanol, or diesel are made from biosourced materials.

References

[1] Green Biologics. Website. (Accessed 6 September 2017, as: http://www.greenbiologics.com/).
[2] Clean Air Act. Website. (Accessed 12 February 2024, as: https://www.epa.gov//clean-air-act-overview).
[3] U.S. Department of Energy, Alternative Fuels Data Center. Website. (Accessed 12 February 2024, as: http://www.afdc.energy.gov/).
[4] US Patent 1315585, Production of acetone and alcohol by bacteriological processes, 1916.
[5] US Patent 4520104, Production of butanol by a continuous fermentation process, 1985.
[6] Renewable Energy World. Syntec & EERC Developing Biomass-to-Butanol Production Technology. (Accessed 30 March 2024, as: renewableenergyworld.com/?s=bio-butanol).
[7] Green Car Congress. Website. (Accessed 30 March 2024, as: https://www.greencarcongress.com/bio butanol/).

[8]   BP. BP and DuPont joint venture, Butamax ®, announces next step in commercialization of bio-
        isobutanol with acquisition of ethanol facility in Kansas. Website. (Accessed 30 March 2024, as:
        https://www.bp.com/en/global/corporate/news-and-insights/press-releases/bp-and-dupont-joint-
        venture.html).
[9]   Gevo. Website. (Accessed 30 March 2024, as: https://gevo.com/).
[10]   Michigan State University, MSU Extension. Biobutanol: A Renewable Fuel for the Future. Website.
        (Accessed 30 March 2024, as: https://www.canr.msu.edu/news/biobutanol_a_renewable_fuel_for_
        the_future).

6 Biofuels from animal and vegetable waste

6.1 Introduction

Modern society uses a wide variety of oils, greases, and fats, and uses them in very many ways. Perhaps an obvious use is cooking oil in restaurant fryers. This oil is used to cook food, but after it has been used repeatedly, becomes contaminated with macroscopic food particles, as well as with smaller-sized residues (think of the smell or aroma of a fryer that has been used exclusively for frying onion rings). When such oils and fats are spent, traditionally they have been discarded. In some cases, this meant simply placing them in the conventional garbage and having them hauled away to be landfilled. In roughly the past two decades, recycling firms have become interested in using such material to produce biofuels [1].

There are certainly other organic waste materials besides cooking oil that have found use in producing some useable biofuel. We discuss here the production of biogas, and biodiesel, the latter from both animal and vegetable sources.

The use of biofuels, specifically biodiesel, has now become established enough that there is an informal nomenclature associated with it. For example, B2 indicates 2% biodiesel mixed with 98% petrodiesel. Likewise, B20 is a mixture, now of 20% biodiesel and 80% petrodiesel. And, as might be expected, B100 is entirely biodiesel.

6.2 Conversion of waste to biogas

The realization that biofuels can be made on a large scale using waste materials of some form is now also established enough that trade organizations exist to promote such use [2–5]. While different organizations have slightly different definitions of the process, the Anaerobic Digestion and Bioresources Association, in Great Britain, states rather straightforwardly at its website:

> Anaerobic digestion is the simple natural breakdown of organic matter into biogas (carbon dioxide and methane) and organic fertilizer called digestate. It is a similar process to that which takes place in the stomach of a cow. [2]

Such statements make the process easy for the general public to understand. They do not, however, connect the current desire to produce biogas to its roots, namely the reduction of pollution. Most biofuels are produced as an alternative to a traditional, petroleum-based fuel. Biogas is often made as a way to reduce the mass and volume of waste material that would otherwise be discarded.

The production of biogas is generally through the anaerobic degradation of any organic matter – with the absence of oxygen being a key element – to yield methane

https://doi.org/10.1515/9783111330259-006

(CH$_4$). The feedstock for biogas is essentially any organic matter that can be composted: farm residues, food waste, sewage, manure, or any other organic waste material. When such material is landfilled in a lined landfill, vents for the biogas can be inserted into the landfill at various points, and the resulting gas collected as it forms. This is sometimes referred to as landfill gas. Figure 6.1 shows a simplified flow diagram of how biogas is produced.

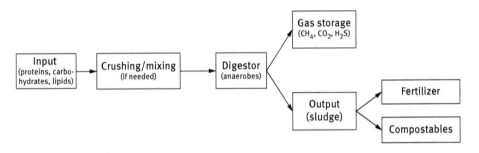

Figure 6.1: Biogas production.

Note that biogas is not the only product in the degradation of waste vegetable and animal matter. The production of organic fertilizer is an important by-product that can also have significant value when sold. This can be in the form of a liquid–solid sludge, or a dry solid, depending on steps in its treatment before it is isolated.

While biogas can be produced in landfill and other open areas, it can sometimes be produced in what is called an anaerobic digester, which is essentially a controlled, closed environment in which an array of microorganisms is used to break down organic materials – usually wastes – in an oxygen-free environment. This is essentially the heart of the operation in Figure 6.1, because a controlled environment means specific anaerobes can be introduced to the material that is to be broken down.

Figure 6.2 shows a basic design of an anaerobic digester, although there are several variations to the design that can be used with essentially equal results. For example, Figure 6.2 does not include mechanical stirrers, but large digesters use them to ensure all the material is exposed to any active microorganisms. As well, some digesters have maintenance hatches, in the event that the container needs to be emptied of the solid digestate that has accumulated (for use as fertilizer) and cleaned. Also, the biogas that is produced is collected at the outlet, and often compressed.

After biogas is produced, it must sometimes have traces of hydrogen sulfide (H$_2$S) removed from it. The most common way this and other contaminants are removed is with water. High-pressure biogas is pumped into a column against flowing water, and the impurities are trapped in the water, and thus removed from the biogas. The end result can be referred to as biomethane. In the flow diagram shown in Figure 6.1, this would represent a further step beyond the gas storage.

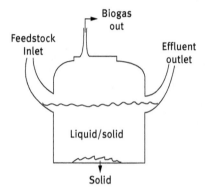

Figure 6.2: Anaerobic digester.

6.3 Use of animal renderings

The production of biodiesel has often centered on some plant as a feedstock for two reasons. First, plants often produce extremely pure triglycerides, without much variation at the molecular level. Animal fats often have differing molecular weight glycerides as part of their triglyceride composition. Second, the industries that supply all forms of meat for consumer consumption and for pet food are mature ones that have put a great deal of thought into the use of all parts of an animal, so that there is very little waste. Fats have traditionally been rendered into soaps and cosmetic ingredients.

Figure 6.3 illustrates the basic chemistry that is involved in converting a triglyceride into 3 mol of what is called a fatty acid methyl ester (FAME), and 1 mol of glycerin. The desired product, the FAME, is a usable fuel. The coproduction of glycerin has driven the price of this material down in recent years, as there is now more than what is needed for all industrial applications.

An example of the use of animal fats to produce biodiesel that has become famous in the past decade is the Perdue Company turning leftover chicken fat into the fuel. Information and press releases describing and discussing this praised the use of this starting material as a new and innovative use of what might otherwise be a waste product [6]. But other animal sources can be used as well, such as beef tallow.

6.4 Use of vegetable oils

Waste vegetable oil has in the recent past been gathered from commercial ventures, as have other types of waste oils, in order to convert this used product to biodiesel.

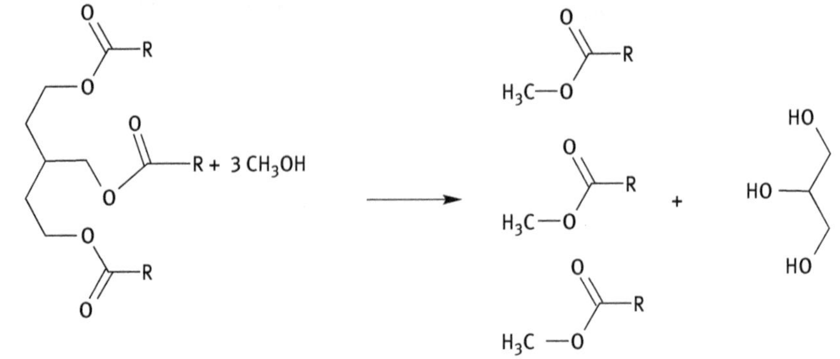

R = alkyl group

Figure 6.3: Transesterification of fats to FAME.

Restaurants that use oil baths for frying food must discard oils after a specified time and repeat number of uses, usually as part of a local or state consumer code involving food safety. Thus, this changing of old for new cooking oil provides a feedstock for the production of biofuel.

The waste oil is often already some long-chain hydrocarbon, but not a triglyceride, so it does not need to undergo a transesterification to make it a usable fuel. It can often be burned as is. And while such a feedstock is not used in automotive transport, it has for over 30 years been used in Germany and other parts of Europe as part of the mixed feed that goes into furnaces to generate heat for residential and commercial buildings.

6.5 Comparisons of biofuels to traditional sources

Fuels such as biodiesel are routinely touted as being a better fuel than their petro-counterparts, simply because they are made from renewable stock. But biodiesel, for example, also burns more cleanly than traditional petrodiesel. It has none of the contaminants associated with petrodiesel, such as sulfur oxides and matter that is particulate in nature. This is a matter of concern to virtually all producers [7–11].

Detractors of the use of vegetable and animal waste products for conversion into biobased fuels point out the sheer scope of the use of automotive fuels, and how little difference it will make against the total if waste vegetable and animal materials were used as fuels. In short, they point out there is simply not enough animal and vegetable waste material to substitute for petroleum-based fuel sources. While this appears to be true, it still seems wiser to utilize these waste materials than to simply discard them in some less useful fashion.

References

[1]   U.S. Environmental Protection Agency. Biodiesel: Fat to Fuel. Website. (Accessed 30 March 2024, as: 19january2017snapshot.epa.gov/www3/region9/waste/biodiesel/index.html).

[2]   Anaerobic Digestion and Bioresources Association. Website. (Accessed 30 March 2024, as: http://ad bioresources.org/).

[3]   American Biogas Council. Website. (Accessed 30 March 2024, as: https://www.americanbiogascoun cil.org/).

[4]   Digestate Certification. Website. (Accessed 30 March 2024, as: http://digestate.org/).

[5]   European Biomass Industry Association. Website. (Accessed 30 March 2024, as: http://www.eubia. org/cms/wiki-biomass/anaerobic-digestion/).

[6]   The Washington Post. Animal Fats Touted As Future Fuel Source. Website. (Accessed 30 March 2024, as: http://www.washingtonpost.com/wp-dyn/content/article/2007/01/02/AR2007010200539_ pf.html).

[7]   Clean Fuels Alliance America. Website. (Accessed 24 April 2024, as: https://www.cleanfuels.org).

[8]   European Biodiesel Board. Website. (Accessed 24 April 2024, as: http://www.ebb-eu.org/).

[9]   Advanced Biofuels Canada. Website. (Accessed 24 April 2024, as: https://advancedbiofuels.ca/).

[10]   Biofuels Association of Australia. Website. (Accessed 24 April 2024, as: http://biofuelsassociation. com.au/).

[11]   Southern African Bioenergy Association. Website. (Accessed 18 January 2018, as: https://www.bi ofpr.com).

7 Yeasts

7.1 Introduction, traditional yeasts

Yeasts have been used by humankind for millennia without a particularly deep understanding of what they are. Their use has normally been a matter of trial-and-error observation for the production of beer, wine, and bread. Archaeologists still debate which is the oldest process that requires some form of yeast – beer and wine production, bread production, or the manufacture of cheese.

Today the many types of yeast can be divided in different ways. In general, yeasts used for beverages include brewer's and wine yeasts. Baker's yeast is that used for breads, and is arguably the most common, or at least the type of yeast known by most consumers. Feed yeasts and bioethanol yeast are two further categories in this larger area. These are produced in large enough amounts that organizations exist dedicated to the promotion of the uses of yeasts [1–4].

7.1.1 Bread yeasts

Saccharomyces cerevisiae (*S. cerevisiae*) – sometimes called baker's yeast – is a yeast species that apparently has been used since ancient times. While it can be airborne, it often exists on the skin of various types of grapes, which has throughout history made it very valuable in wine production. Yet it is also crucial to bread baking.

The end result of the addition of baker's yeast to dough is quite obvious: it leavens bread. It does so by converting sugars in the dough to both carbon dioxide and ethanol. The reaction can be shown in simplified form in Figure 7.1, although the amounts of one product or the other can be adjusted by variations in the reaction conditions.

$$C_6H_{12}O_6 \rightarrow 2\,C_2H_5OH + 2\,CO_{2(g)}$$

Figure 7.1: Baker's yeast fermentation.

The large-scale production of baker's yeast has in the past century become a specialized industry. A few large firms dominate the production, although there are smaller firms that do produce baker's yeast and other, sometimes more specialized yeasts [5–14]. Table 7.1 shows the largest companies in terms of what they produce.

The production of yeast itself is through a modification of the conditions under which yeast is used to produce bread (or other foods). Aerobic fermentation under controlled conditions can produce more yeast biomass, and significantly less alcohol. What are called mother yeasts or more commonly mother cultures are kept by manufacturers, and when used are placed in a sugar solution that is rich in vitamins, minerals, and nitrogen. The life cycle of yeast is quick enough that several generations

https://doi.org/10.1515/9783111330259-007

Table 7.1: Yeast producing corporations.

Name	Product(s)	Comments	Web site
AB Vista	Markets "Vistacell"	Major producer of animal feed enzymes	https://www.abvista.com
AB Mauri	"Fleischman's" yeast products		http://abmna.com/
Angel Yeast Co., Ltd.		In China and Singapore	http://en.angelyeast.com/
DSM		Wide range of products, including food supplements	http://www.dsm.com/markets/foodandbeverages/en_US/home.html
GB Plange		Acquired by AB Mauri in 2014	
Lallemand		Producing yeasts since 1923	http://www.lallemand.com/
Lesaffre Group		Products in healthcare as well	http://www.lesaffre.com/
Wyeast	Markets multiple brewing/baking yeasts		http://www.wyeastlab.com/yeast-fundamentals

Source: [5–14]

can be produced, then separated from the host liquid. This material can be sold as a form simply called "liquid cream yeast," or can be dried and sold as dry yeast. The general public tends to think of yeast as small packets of dried material – and indeed, grocery stores do sell yeast as such for consumer baking needs. Figure 7.2 shows a display of different company's yeasts for sale directly to consumers. But it can be filtered, dried, and sold as multikilogram bags. Such sales are to businesses, such as bakeries, that require large amounts of yeast for the large-scale production of bread. The production steps are illustrated in Figure 7.3.

Concerning the steps in yeast production, more detail than this often involves proprietary steps on the part of manufacturers. Mother cultures are often kept much like a corporate trade secret, and access to them can be tightly controlled.

7.1.2 Beer and wine yeasts

Throughout most of recorded history, both beer and wine have been produced using airborne, wild yeasts. Indeed, the very existence of yeast was unknown until relatively modern times, as exemplified in the German Reinheitsgebot – the Purity Law –

Figure 7.2: Commercially available yeasts.

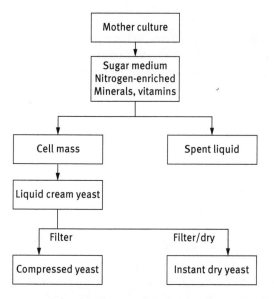

Figure 7.3: Yeast production.

of 1516, which spelled out what ingredients were allowed in German beers of the time. The section of it specifically dealing with ingredients states:

> We wish to emphasize that in future in all cities, markets and in the country, the only ingredients used for the brewing of beer must be Barley, Hops and Water. Whosoever knowingly disregards or transgresses upon this ordinance, shall be punished by the Court authorities' confiscating such barrels of beer, without fail. [15]

This law, one of the oldest still on record that deal with ingredients for a specific food or beverage, does not mention yeast because it was unknown at the time. And even today, the addition of other ingredients, such as rice, is not permitted in numerous German and other European beers.

For home brewers and home vintners today, yeast can be purchased from specialty shops, along with all the other equipment needed to produce beer and wine in what can be called a cottage industry scale. Larger corporate concerns tend to use yeasts that are proprietary to their companies. Curiously, one large producer in the United States announced in May 2018 that it would produce a beer that was as close as possible to the recipe used by General, later President, George Washington, indicating it is possible to scale up a recipe that had in the past been used on a smaller scale. The announcement did not specify whether the yeast to be used was the same or in some way related to that used in Washington's time [16].

7.2 Recent yeast modifications

Among yeasts, *S. cerevisiae* has been modified several times, in large part because it is a simple organism that is readily adaptable to a wide variety of conditions, including those of a wide-ranging pH [17]. It has been engineered to produce the following, which is a nonexhaustive list:
- Arsenic bioremediation
- Bioethanol
- Human serum albumin
- Insulin
- Isoprenoids
- Various vaccines
- Xylose

The full potential of yeast in producing pharmaceuticals has probably not yet been harnessed, since a significant number of what may be called biopharmaceuticals produced using yeasts today were produced by more traditional methods in past decades. There are thus probably more reactions that are still being performed in a traditional manner that might be adapted to yeast as a production method [18]. As with virtually all uses of yeast, these changes are because the resulting method is economically more feasible than that which came before it.

References

[1]   Confederation of European Yeast Producers. Website. (Accessed 24 April 2024, as: https://www.cofa lec.com/).
[2]   The World of Yeast, Biology. Website. (Accessed 24 April 2024, as: https://www.cofalec.com/the-world-of-yeast/biology/).

[3]   World Yeast Congress. Website. (Accessed 24 April 2024 as: https://yeast.conferenceseries.com/).

[4]   Yeast Market. Website. (Accessed 24 April 2024, as: https://www.marketresearch.com/yeast-mar
ket.html).

[5]   AB Vista. Website. (Accessed 24 April 2024, as: https://www.abvista.com).

[6]   AB Mauri. Website. (Accessed 24 April 2024, as: http://abmna.com/).

[7]   Angel Yeast Co., Ltd. Website. (Accessed 24 April 2024, as: http://en.angelyeast.com/).

[8]   DSM. Website. (Accessed 24 April 2024, as: https://www.dsm.com/food-beverages/en_US/
home.html#).

[9]   Lesaffre Group. Website. (Accessed 24 April 2024, as: http://www.lesaffrestore.com/).

[10]   Lallemand. Website. (Accessed 24 April 2024, as: http://www.lallemand.com/).

[11]   Explore Yeast. Website. (Accessed 24 April 2024, as: http://www.exploreyeast.com/).

[12]   Manufacturing of Nutritional Yeast: National Emission Standards for Hazardous Air Pollutants
(NESHAP). Website. (Accessed 24 April 2024, as: epa.gov/stationary-sources-air-pollution/
manufacturing-traditional-yeast-national-emission-standards).

[13]   Commercial Yeast Production. Website. (Accessed 24 April 2024, as: http://www.lallemand.com/en).

[14]   Wyeast Website. (Accessed 24 April 2024, as: http://www.wyeastlab.com).

[15]   Germany's Purity Law. Website. (Accessed 24 April 2024, as: http://www.brewery.org/library/Rein
Heit.html).

[16]   George Washington's Handwritten Beer Recipe. Website. (Accessed 24 April 2024, as: bonappetit.
com/drinks/beer/article/George-washington-s-handwritten-beer-recipe).

[17]   Yeast: Love and Fear, Death and Beer – Single-celled fungi all around us do so much good and so
much bad. The Atlantic, Website. (Accessed 24 April 2024, as: https://www.theatlantic.com/health/
archive/2013/05/yeast-love-and-fear-death-and-beer/275786/).

[18]   Nielsen, J. Production of biopharmaceutical proteins by yeast, Advances through metabolic
engineering, Bioengineered, 4(4), 207–211, 2013.

8 Vitamins

8.1 Introduction

Vitamins – called so because it was originally thought that all of these substances contained nitrogen atoms in the form of amines, hence the name "vita-amines" – are a group of relatively small molecules termed "essential," because the human body is incapable of producing them, and therefore they must be consumed. The earliest reports of any vitamin are credited to Kazimierz Funk (often, Casimir Funk), who in 1912 published the book, *The Vitamines* [1]. Throughout history, small subsets of people have suffered from one vitamin deficiency or another, sometimes with disastrous, fatal results.

The stories that accompany the discovery of several vitamins are colorful, and involve some excellent human minds solving problems that ultimately saved an enormous number of human lives. Perhaps the best known example is the story of the discovery of vitamin C by British naval surgeon James Lind, and the subsequent eradication of scurvy in sailors of the British Navy, based on enforcing consumption of a daily ration of lime juice for each sailor when at sea. Lind's saga, and those of others, was normally not one of synthesis of a particular vitamin, but rather a quest to find some food already in nature that provides humans with the necessary source.

Currently, thousands of tons of vitamins are produced annually, which seems like an enormous amount when we measure this against the individual bottles sold to the consumer in drug stores, pharmacies, and health food stores. But a large percentage of the vitamins produced today are not sold for human consumption; rather, these vitamins are used in animal feed. Such uses tend to keep animals healthy as they grow, and in some cases enhance their size and weight, which is important when they are sent to slaughter houses. Overall, vitamins have become a large enough set of products, with a large enough market, that there are several trade organizations devoted to their sale and proper use [2–11].

There are currently 13 recognized vitamins, and additionally several further molecules that are sometimes called "vitamin-like substances," the full understanding of which is a continuing field of study. Some vitamins are still extracted from natural sources, while others can be synthesized starting from simple, usually organic, molecules. In the latter case, the ultimate starting material may be some molecules derived from petroleum. There are some vitamins though that require what can be called a biotechnological step (or more than one) to ensure their production occurs in high yield, and sometimes with correct stereochemical configuration. It is this latter class of vitamins that we will discuss in this chapter.

https://doi.org/10.1515/9783111330259-008

8.2 Vitamin B$_2$, riboflavin

Vitamin B$_2$ is oftentimes still referred to as riboflavin, and is a water-soluble vitamin. While it is found naturally in a rather wide variety of animal and vegetable sources, such as meats and cheeses, it is liver and kidney that are animal sources high in the vitamin. Curiously, certain yeasts, as well as almonds and mushrooms can also be high in riboflavin. The Lewis structure is shown in Figure 8.1.

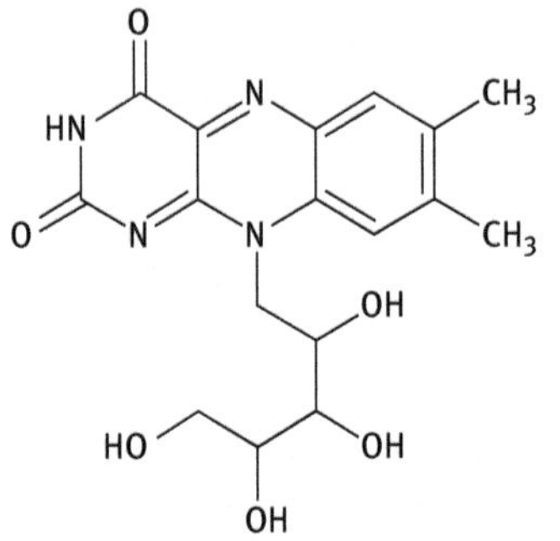

Figure 8.1: Riboflavin structure.

Vitamin B$_2$ is now produced on a large, industrial scale with several different microbial systems and enzymes. Several companies have interests in the production of this vitamin, and different companies choose to use different organisms, for economic reasons. For example, BASF has utilized *Ashbya gossypii* extensively, and with it produces riboflavin for human consumption as well as for an animal feed additive [12, 13].

The use of different organisms to produce riboflavin is now several decades old, which Stahmann et al. reported in 2000. They state:

> Three microorganisms are currently in use for industrial riboflavin production. The hemiascomycetes *Ashbya gossypii*, a filamentous fungus, and *Candida famata*, a yeast, are naturally occurring overproducers of this vitamin. To obtain riboflavin production with the Gram-positive bacterium *Bacillus subtilis* requires at least the deregulation of purine synthesis and a mutation in a flavokinase/FAD-synthetase [13].

8.3 Vitamin B$_3$, niacin

Vitamin B$_3$, still often referred to simply as niacin, is another vitamin that possesses enough polar portions and is classified as water soluble. While it is found in a rather broad variety of animal and vegetable sources, there is still a need for large enough quantities of it that large-scale industrial processes are used for its production. Figure 8.2 shows the Lewis structure for it.

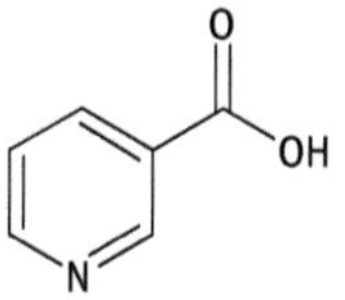

Figure 8.2: Molecular structure of niacin.

Since it is a relatively simple organic molecule, niacin can be manufactured from 3-methylpyridine, which in turn is produced from acrolein and ammonia. Such methods are classified as traditional organic synthesis. But it is very often produced enzymatically, from tryptophan, as illustrated in Figure 8.3. With all processes combined, each year almost 10 million tons of this vitamin are manufactured. More than half of this is designated for and consumed in various animal feeds.

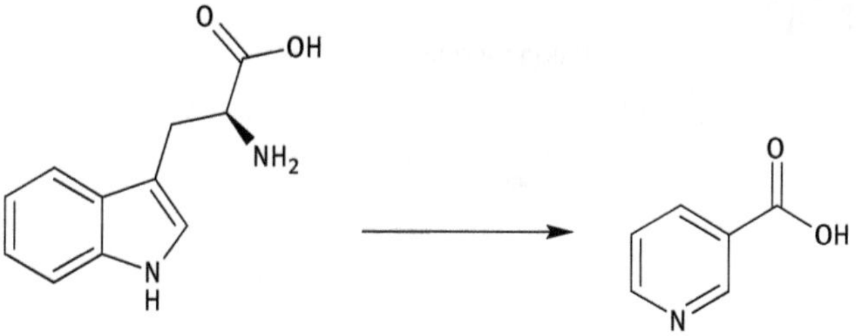

Figure 8.3: Niacin from tryptophan.

What is shown in Figure 8.3 is a simplification of the reactions effected by indoleamine 2,3-dioxygenase and formamidase, the enzymes needed ultimately to produce niacin in a multistep process. But for our purposes, the starting material and end product are important, as opposed to each intermediate transformation and mechanistic step.

Recently, the personal consumption of more niacin per day than is considered necessary has been linked to lower cholesterol and triglyceride levels, thus enhancing the market for it in health food stores. Figure 8.4 shows a bottle of niacin, and indicates it is "flush free." This means that the side effect known as flushing – itchiness, warmth, and tingling of the skin – does not occur with this brand of niacin capsules.

8.4 Vitamin B$_6$

Vitamin B$_6$ is also known as pyridoxine, and is another that is classified as a water-soluble vitamin. While the vitamin can exist in seven differing forms, the Lewis structure of the active form is shown in Figure 8.5. This form is often called pyridoxal phosphate, and is abbreviated PLP.

Figure 8.4: Niacin capsules.

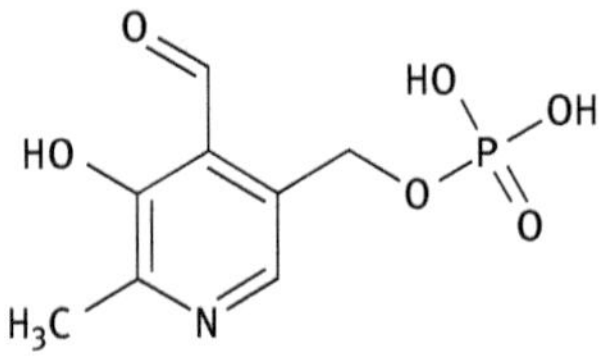

Figure 8.5: Structure of vitamin B_6.

The large-scale production of vitamin B_6 is another that requires enzymatic and microbial processes. Different companies employ different routes for its production. Companies such as Takeda and Daiichi each employ methods unique to their firm, and keep details of their processes proprietary [14, 15].

Vitamin B_6 has been shown to promote health of the human heart; and there now appears to be a connection between its consumption and a reduced chance of stroke and other cardiovascular diseases. Because of this, health and nutrition stores now market vitamin B_6 as being heart healthy, as shown in Figure 8.6.

8.5 Vitamin B_{12}

Vitamin B_{12} or cobalamin is the most complex of all the vitamins in terms of its structure. Although its synthesis via traditional organic chemistry has been accomplished, the length and complexity of such proved that it would be impossible to enlarge this type of synthesis to an industrial scale.

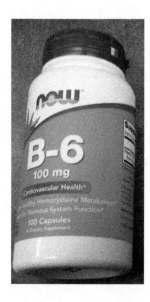

Figure 8.6: Vitamin B$_6$ capsules.

Routinely, the industrial-scale production of vitamin B$_{12}$ is accomplished via a fermentation using *Pseudomonas denitrificans,* a form of bacteria that can be nurtured and controlled to optimize yields.

Figure 8.7 shows the Lewis structure of vitamin B$_{12}$. Note the cobalt atom in the center of the structure.

R = 5′-deoxyadenosyl, Me, OH, CN   Figure 8.7: Structure of vitamin B$_{12}$.

Claims have been made that numerous bacteria can produce vitamin B_{12}, although only a few are used industrially to perform almost all production. The following have been recorded as having been used in vitamin B_{12} production:

- Acetobacterium
- Aerobacter
- Agrobacterium
- Alcaligenes
- Azotobacter
- Bacillus
- Clostridium
- Corynebacterium
- Flavobacterium
- Lactobacillus
- Micromonospora
- Mycobacterium
- Nocardia
- Propionibacterium shermanii
- Protaminobacter
- Proteus
- Pseudomonas denitrificans
- Rhizobium
- Salmonella typhimurium
- Serratia
- Streptococcus
- Streptomyces griseus
- Xanthomonas

The use of *Streptomyces griseus* for vitamin B_{12} was prevalent for many years, but this has been displaced by the just-mentioned *P. dentrificans* and *P. shermanii* to a large extent. Companies involved in the large-scale production of this vitamin, such as Sanofi [16], tend to keep their organisms, and the conditions under which they are employed, as proprietary information.

Vitamin B_{12} has been associated with numerous health benefits, although the only one that appears to have been proven is the promotion of red blood cell formation. Because of the perceived health benefits, however, vitamin B_{12} is sold in health and nutrition stores, usually in amounts far in excess of the believed daily requirement. Figure 8.8 shows an example of this. Note how much higher than the 100% recommended daily dosage each capsule contains; and further note at the base of the figure that the producer makes no guarantee of the health benefits of the vitamin.

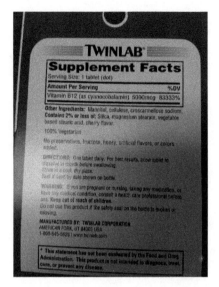

Figure 8.8: Label on a packet of vitamin B_{12}.

8.6 Vitamin C

Vitamin C is a well-known and extensively studied nutrient in such foods as citrus fruits, in breakfast staples such as orange juice, and the disease associated with its deficiency – scurvy – has a long and highly studied history. Currently, a person can ingest a diverse enough diet that they take in enough vitamin C without the need for supplements. When vitamin C is produced on a large scale, however, it requires the use of *Acetobacter*, as illustrated in Figure 8.9.

Note that each step does not require some enzyme or other biotechnological means for a transformation. However, no more economically efficient manner has yet been devised to effect the reduction of the alcohol to a ketone that is required before the oxidation of one of the terminal alcohols to a carboxylic acid. Without this use of *Acetobacter*, a racemic mixture is produced, and thus a pure product does not form quantitatively.

Perhaps obviously, there are other vitamins than those discussed here, which can be produced on an industrial scale. But these are those for which some biotechnological step or process is required in the overall synthetic process.

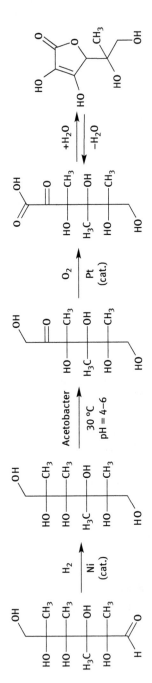

Figure 8.9: Vitamin C production.

References

[1] Casimir Funk. The Vitamines. Classic Reprints, ISBN: 978-1332308750.
[2] BIO. Biotechnology Innovation Organization. Website. (Accessed 24 April 2028, as: https://www.bio.org).
[3] Natural Products Association. Website. (Accessed 24 April 2024, as: https://www.npanational.org/).
[4] US Food and Drug Administration. Website. (Accessed 24 April 2024, as: https://www.fda.gov/).
[5] European Medicines Agency. Website. (Accessed 24 April 2024, as: https://www.ema.europa.eu).
[6] Australian Government Therapeutic Goods Administration. Website. (Accessed 24 April 2024, as: https://www.tga.gov.au/).
[7] U.K. Food Standards Agency. Website. (Accessed 24 April 2024, as: https://www.food.gov.uk/).
[8] Canadian Food Inspection Agency. Website. (Accessed 24 April 2024, as: https://inspection.canada.ca).
[9] International Food Additives Council. Website. (Accessed 24 April, 2024, as: https://www.foodingredientfacts.org/about-us).
[10] AAHSA. Health Food and Supplements Association. Website. (Accessed 24 April 2024, as: https://aahsa.org.sg/hfsa/?standard).
[11] Food Additives and Ingredients Association. Website. (Accessed 24 April 2024, as: http://www.faia.org.uk/).
[12] BASF. Website. (Accessed 24 April 2024, as: www.basf.com).
[13] Stahmann, K.P., Revuelta, J.L., and Seulberger, H. Three biotechnical processes using Ashbya gossypii, Candida famata, or Bacillus subtilis compete with chemical riboflavin production. Applied Microbiology and Biotechnology, May 2000, 53(5), 509–516. (https://link.springer.com/article/10.1007%2Fs002530051649).
[14] Daiichi. Website. (Accessed 24 April 2024, as: https://www.daiichisankyo.com/about_us/).
[15] Takeda. Website. (Accessed 24 April 2024, as: https://www.takeda.com).
[16] Sanofi. Website. (Accessed 24 April 2024, as: http://www.sanofi.com/en/).

9 Pharma/Drugs

9.1 Introduction

The production of pharmaceuticals using some biological agent in their synthesis, some biotechnological step as it were, has a long history, although much of that history involves nothing more than cultivating certain plants that were known to have a medicinal value, and selecting for one characteristic throughout several generations of growth. One can argue that the understanding of the role of some biological component in the synthesis of early drugs is one of our oldest connections between biology and chemistry, with only the arts of brewing, baking, and cheesemaking processes predating it. Many of the earliest drugs were simply some plant-based or animal-based material that might be boiled, dried, or in some other way concentrated before actual use. Perhaps the biggest of the classic examples is the use of willow tree bark as a chewable pain reliever.

In the last century, there has been an explosion of drugs that have been brought to the market, and many of them require some biotechnological step or steps in their large-scale synthesis due to the complexity of the target molecule. In many cases, some form of recombinant DNA is now used to express a material, such as a protein, for use either as a drug or as a precursor to a pharmaceutical. We have not chosen to present a detailed discussion of each. Rather, we have presented several select, prominent pharmaceutical materials, and created a table discussing those that have been produced in the largest amounts.

9.2 Insulin

The treatment of diabetes has a relatively long history, at least as far as some form of drug treatment is concerned. The pancreatic islets or Isles of Langerhans were discovered in 1869 as part of the pancreas, as an endocrine producer, by Paul Langerhan. This is where healthy humans normally produce insulin in vivo.

When a human is no longer able to produce insulin, some nonhuman source must be found to supply it. The production of insulin advanced to the extraction of insulin from various animals, often pigs and cows, and was used until the 1980s as the main treatment for diabetes, after extraction and purification [1]. The insulin of both animals is extremely close to that produced by human beings. Indeed, prescription animal insulin is still available on the market today, although not widely. The drug name "hypurin," for example, is that used for animal insulin produced by Wockhardt UK [2].

Even though animal-derived insulin was the major insulin product for years, its amino acid sequence and structure was first determined by Dr. Frederick Sanger in 1951 and 1952. For his work, Sanger would win the 1958 Nobel Prize in chemistry (and by win-

https://doi.org/10.1515/9783111330259-009

ning a share of a second in 1980 for his work on base sequences and nucleic acids remains one of only a very few people to have won two, in any category).

What can be called the first synthetic insulin was produced in the 1960s – although on a lab scale, not a commercial scale. Commercial production would require another two decades, but has since become a staple drug for several pharmaceutical companies.

The amino acid sequence of bovine insulin is shown in Figure 9.1.

What is generally called human insulin, or recombinant insulin, can be produced using either the yeast *Saccharomyces cerevisiae*, or *Escherichia coli*, as the basis for the expression of the protein. This was first attempted in the late 1970s, and the first synthetic human insulin was made commercially available in 1982. Since this time, Eli Lilley has sold such insulin under the name Humulin. Since then, several companies have developed different synthetic insulin products, and continue to refine their offerings. Table 9.1 is a nonexhaustive list of commercial insulin producers.

Insulin can be produced by expressing one chain independently of the other, but such a synthesis suffers from any problems that might occur when the two chains are recombined. If recombination is not optimal, the end product does not function effectively.

Table 9.1: Companies that produce insulin.

Company name	Insulin name	Comments
Amylin	Byetta	Company acquired by Bristol-Myers Sqibb and AstraZeneca
Amylin	Bydureon	
Eli Lilley	Humulin	
MannKind	Afrezza	Requires FDA approval
Novo Nordisk	NovoLog pen, Levemir, Victoza	
Pfizer	Exubera	Inhalable form
Sanofi	Lantus, Aprida	

Source: [3–8].

9.3 Taxol (paclitaxel)

This anticancer drug was the subject of an intense debate in the 1990s, specifically about how to manufacture it in large enough quantities to meet the growing market demand. The natural source had been the bark of the Pacific yew tree (*Taxus brevifola*), and pro-

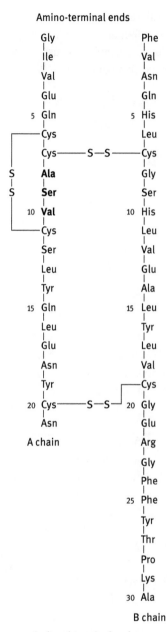

Amino-terminal ends

Carboxyl-terminal ends

Figure 9.1: Amino acid sequence of bovine insulin.

jected needs for the compound were such that there were simply not enough trees in existence to meet the need even if they were harvested to extinction. Thus, some other route to the product was required.

While several prominent research groups devoted considerable time to the total synthesis of taxol, and while more than one total synthesis was found, it was ultimately a biotechnological method that was used to scale up production, simply because of the complexity of the molecule, which includes 11 stereocenters, as shown in Figure 9.2.

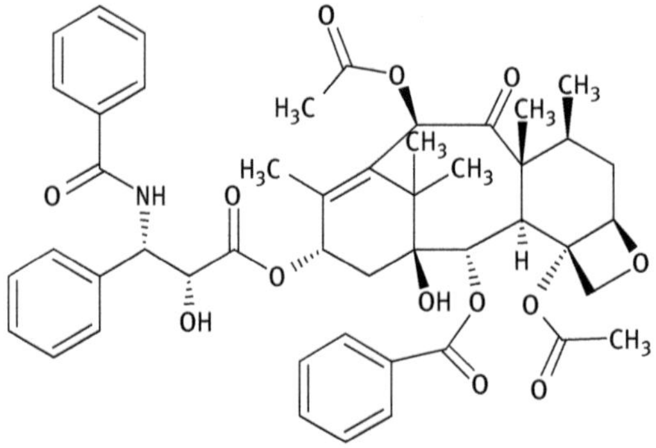

Figure 9.2: Molecular structure of taxol.

Most recently, a form of plant cell fermentation is utilized in the production of taxol, requiring a specific cell line, but one that is generated using *Penicillium raistrickii* fungus in fermentation tanks. This removes the need for any yew tree as a source, and produces a product that needs nothing further than chromatographic purification and crystallization. Compared to the original synthetic sequences for taxol production, this method is far more economical in terms of both energy expenditure and reagents or solvents used. This method has been championed by Phyton Biotech, Inc., using proprietary techniques, and which advertises itself as "Global Leaders in Plant Cell Fermentation [9]."

Several names exist for marketed versions of this drug, including Abraxane, Taxol, Onxol, Placlitaxel Novaplus, PremierPro RX, RX PACLitaxel, and Nov-Onxol.

9.4 Interferons

The use of interferons, such as IFN-a$_2$ to treat hepatitis as well as multiple sclerosis (MS), is another example of the production of a drug, or drugs, that requires recombinant DNA for its synthesis on a large scale [10].

Interferons are a group of proteins that the body produces to alert cellular machinery to the presence of pathogens. Thus, there are interferons present in all humans, but only in tiny amounts.

The original production of usable amounts of interferon from blood donors – the first step in determining what was needed to market interferons as a drug – was a massive effort. Over 90,000 donors were required to concentrate merely 1 g of the material [11]. Current efforts circumvent the use of donors as a source. Current methods use recombinant DNA, but remain proprietary to specific companies [11–13].

9.5 Other common drugs

As mentioned, there are numerous drugs that require some biotechnological step or steps in their synthesis. Table 9.2 is a nonexhaustive list of several of them, with the caveat that this table will undoubtedly change as time passes, as the public demand for a specific drug waxes or wanes. In a large number of cases, the key breakthrough in the production of any of these drugs is finding some organism that can be modified with a piece of DNA to express either the drug or some precursor that can then be used to produce the target. As seen from the discussion in previous sections, the use of simple organisms such as *E. coli* or *S. cerevisiae* tends to be preferred, because it is not particularly difficult to effect specific modifications in them, and because their life cycle is so rapid.

Table 9.2: Examples of other drugs produced via some biotechnological route.

Name	a.ka.	Used for	Organism needed	Company
Aranesp	Darbepoetin	Rheumatoid arthritis (RA)	Tumor necrosis factor (TNF)	Amgen
Avastin	Bevacizumab	Antiangiogenesis, cancer		Genetech
Avonex	Interferon beta-1a	MS		Biogen Idec
Betaseron	Interferon beta 1b	MS		Bayer Schering
Copaxone	Glatiramer acetate	Relapsing MS		Teva
Enbrel	Etanercept	Rheumatoid arthritis		Amgen product
Epogen	Epoetin alfa	Anemia, kidney disease		Amgen
Erbitux	Cetuximab	Cancer, head and neck		ImClone

Table 9.2 (continued)

Name	a.ka.	Used for	Organism needed	Company
Herceptin	Trastuzumab	Breast cancer		Genentech
Humalog	Insulin lispro	Diabetes		Eli Lilly
Humira	Adalimumab	Psoriasis		Abbott
Lantus	Insulin glargine	Insulin analog	*E. coli*	Sanofi-aventis
Lucentis	Ranibizumab	Antiangiogenesis, cancer		Genentech
Neulasta	Pegfilgrastim	Anticancer	*E. coli*	Amgen
Prevnar	Diphtheria CRM$_{197}$ protein	Pneumococcal bacterial infection		Wyeth
Procrit/Eprex	Epoetin alfa	Anemia		Ortho Biotech
Rebif	Interferon beta-1a	Relapsing MS		Merck Serono
Remicade	Infliximab	Rheumatoid arthritis	Chimeric monoclonal antibody (mouse, human)	Johnson & Johnson
Rituxan	Rituximab	Cancer, autoimmune disease		Genetech

Source: [14–23].

References

[1] Diabetes.co.uk, The Global Diabetes Community. Website. (Accessed 24 April 2024, as: https://www.diabetes.co.uk/).
[2] Wockhardt, UK. Website. (Accessed 24 April 2024, as: http://www.wockhardt.co.uk/).
[3] Amylin. Website. (Accessed 24 April 2024, as: www.astrazeneca.com/our-therapy-areas/medicines.html).
[4] Eli Lilly. Website. (Accessed 24 April 2024, as: https://www.lilly.com/).
[5] MannKind. (Accessed 24 April 2024, as: www.mannkindcorp.com).
[6] Novo Nordisk. (Accessed 24 April 2024, as: www.novonordisk-us.com).
[7] Pfizer. (Accessed 24 April 2024, as: www.pfizer.com).
[8] Sanofi. Website. (Accessed 24 April 2024, as: https://www.sanofi.com).
[9] Phyton Biotech, Inc. Website. (Accessed 24 April 2024, as: https://phytonbiotech.com/).
[10] Biogen plegridy. Website. (Accessed 24 April 2024, as: https://plegridy.com/?cid=aff-interferons-redirect-hp?).
[11] Buckel, P., ed. Recombinant Protein Drugs (Series: Milestones in Drug Therapy), 2001, ISBN: 3-7643-5904-8.

[12] Biotechnolgy – New Directions in Medicine. F. Hoffman-La Roche, Ltd., Basel, Switzerland, 2008.

[13] Avonex. Website. (Accessed 24 April 2024, as: https://www.avonex.com/).

[14] Abbott. Website. (Accessed 24 April 2024, as: https://www.abbott.com/).

[15] Amgen. Website. (Accessed 24 April 2024, as: https://www.amgen.com/).

[16] Bayer. Website. (Accessed 24 April 2024, as: https://bayer.com/).

[17] Biogen Idec. Website. (Accessed 24 April 2024, as: https://www.biogen.com/).

[18] Genentech. Website. (Accessed 24 April 2024, as: https://www.gene.com/).

[19] ImClone-Eli Lilly. Website. (Accessed 24 April 2024, as: https://www.lilly.com).

[20] Johnson & Johnson. Website. (Accessed 24 April 2024, as: https://www.jnj.com/).

[21] Merck KGaA. Website. (Accessed 24 April 2024, as: https://www.emdgroup.com/en/expertise/health care.html).

[22] Fierce Healthcare. Website. (Accessed 24 April 2024, as: https://www.fiercehealthcare.com/health care/ortho-iotech-oncology-research-development-unites-johnson-johnson-biopharmaceutical).

[23] Wyeth – Pfizer. Website. (Accessed 24 April 2024, as: www.pfizer.com).

10 Insect sources

10.1 Introduction

The diversity of insect life on the Earth is absolutely enormous. Estimates are that there are tens of millions of different insect species on the planet, and that they may constitute more than 90% of nonplant life on the Earth. Throughout history, humans have interacted with insects in both positive and negative ways. The destruction of sown crops by locusts and other swarming insects is certainly one of the most negative interactions between humans and insects. A far more positive one is the husbanding of bees for their production of honey and beeswax. Similarly, the production of silk from silkworms is a human–insect interaction that is hundreds of years old, and that has been remarkably productive for all of that time. Additionally and perhaps obviously, several types of insects have become part of the diet of different peoples.

In the past century, there have been a widening series of interactions between humans and insects, at least in terms of humans overseeing the production of some insect source or product for our own benefit. Insect sources are now used in applications as wide ranging as food colorings and textile dyes.

10.2 Termite gut bacteria and cellulosic ethanol

As discussed in Chapter 4, cellulosic ethanol is that made from plant material that is high in cellulose, and not in starch. The breakdown of cellulose to single-ring sugars has always presented a problem, since linkage cannot be degraded by the yeasts that degrade starch.

Efforts have been made to break down cellulose via the same method by which termites break it down. Interestingly, it is not termites that actually break down cellulose; rather, they ingest wood and it is a gut bacteria that termites also ingest that is capable of breaking what is called the beta linkage from one glucose ring to another. This is the hurdle that must be overcome so that the ultimate transformation from cellulose to ethanol can occur.

Currently, no company has advertised that it is making cellulosic ethanol on an industrial scale through the use of termite gut bacteria. Yet there appear to be several companies and academic researchers who are experimenting with the best conditions to bring this technique to an industrial scale.

https://doi.org/10.1515/9783111330259-010

10.3 Cochineal

The use of cochineal insects has proven to be a profitable way to impart a red color to certain cosmetics and foods. The insect is a very small, scale insect which lives on various cactus species, and when dried and crushed forms a bright red powder (it is of the species *Dactylopius coccus*). The color is largely due to carminic acid; the structure of which is shown in Figure 10.1.

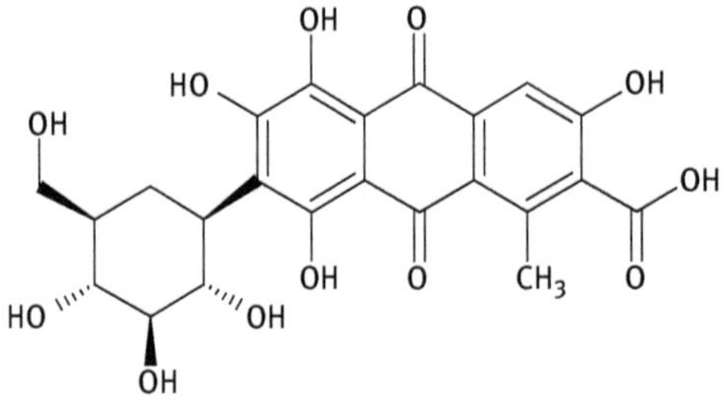

Figure 10.1: Lewis structure of carminic acid.

Cochineal is now produced and used on a large enough scale that it has an identifying number, E120, in the European Union system of colorings. It is used primarily for red dyes, but can produce orange colors under the right conditions, specifically the pH of its solutions [1]. The product is exported from Mexico and Peru, among other nations, and used in wines as well as lipsticks and eye liners to impart a red color. It can be used in a variety of other foods as well, again usually to impart or enhance a red color.

Production of cochineal is through the gathering and drying of the female cochineal insects and the eggs. Estimates are that 80,000–150,000 insects must be dried to produce a kilogram of the final product. Since the insects are gathered from the cactus plants on which they live, clearly the process is a labor-intensive one. Yet the harvesting of cochineal is a mature industry, although it can also be described as a cottage industry, simply one that has been widely undertaken.

10.4 Silk

The production of silk from insects – a variety of species, but generally those simply called silk worms – has an ancient history, and in the past century has undergone significant expansion into a large-scale industry [2–4]. The production of virtually all farmed silk is through mulberry plants, upon which silk worms feed routinely, the species *Bombyx mori*. Large-scale silk production and silk farming now focuses on moth caterpillars, which can feed on mulberry leaves and which can be concentrated in an area where harvesting of the cocoons is economically feasible.

The molecular structure of silk can be complex, but Figure 10.2 shows it in a somewhat simplified form. This repeat structure – glycine-serine-glycine-alanine-glycine-alanine – is largely what forms the beta sheets of proteins that comprise a large portion of silk, and thus what ultimately make silk as strong and resilient as it is when woven into sheets and garments.

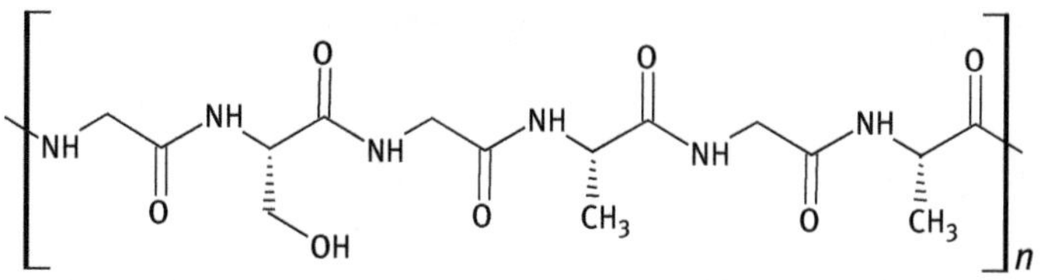

Figure 10.2: Simplified structure of silk.

10.4.1 Silk production

The process of growing silk worms is known as sericulture, from the Latin serica, meaning silk, and cultura, meaning cultivation. Broad steps in its production can be listed as:
1. breeding the worms on mulberry leaves;
2. allowing the worms to pupate in cocoons;
3. harvesting the cocoons;
4. boiling the cocoons in water;
5. extracting the fibers from boiling water;
6. spinning the raw silk on reels.

While this appears to be a straightforward process, in the past there has been some controversy over it. Individuals where the worms are cultivated are sometimes strictly Buddhist, and object to the killing of any form of life, including silk worms. They prefer to allow the worm to chew its way out of the cocoon, rather than to kill it in boiling water while still in the cocoon. Allowing the worm to mature produces silk fibers that are inferior to those produced when the cocoon is boiled, and thus fibers that are less valuable.

In terms of industrial scale, China has led the world in silk production for several years and appears poised to continue to do so.

The general public tends to think of silk as a luxury item for certain pieces of expensive clothing, but there are several other uses as well. A nonexhaustive list includes:
1. clothing – blouses, dresses, pajamas, shirts, ties;
2. medicine – sutures, drug delivery;
3. furniture – bedding, upholstery coverings, wall hangings;
4. industry – gunpowder satchels, parachutes, specialty tires.

Because of silk's biocompatibility as well as biodegradability, it continues to be examined for potential use in a wide variety of medical applications.

The possibility of producing different types of silk on a large scale is of interest because the traditional methods are considered intensive in terms of water and other materials. Fertilizers are consumed in growing mulberry plants, and significant amounts of land must be used to grow enough plant material for the thousands of cocoons that must be harvested to make enough silk for even one application, such as a garment.

The possibility has existed for decades of producing silk from various spider species. The promise is of a material that is even lighter and stronger than existing silks. None have yet been commercially successful however because of the difficulties involved in concentrating and husbanding such arachnids. Certain species are both territorial and cannibalistic in nature. However, in May 2016, the trade magazine *Chemical & Engineering News* briefly reported that Bolt Threads had raised significant amounts of capital for the production of spider silk through a different method. The business report indicated that, "The firm genetically modifies yeast to express silk in industrial fermenters" [5]. Thus, the search for silk based on spiders continue, but utilizing techniques apparently much like those used in ethanol production.

10.5 Beeswax

As mentioned, bees and humans have lived in close contact for millennia, and humans have used both honey and beeswax for as long, the former as a food source, the latter in a wide variety of applications. Today, several regional organizations exist promoting beekeeping and the uses of products derived from bees [6–12]. Beekeeping today involves the growth and husbanding of hives by humans, using frames that make the harvesting of both products and honeycomb economically advantageous.

The Lewis structure of beeswax is not easy to represent because there are variations in waxes, based on the diets of particular bee colonies. Wax is made up of a mixture of long-chain aliphatic esters – the greater fraction – and long-chain alcohols – generally a smaller fraction. In an attempt to represent a simplified structure, Figure 10.3 shows a structure of a well-characterized component of beeswax.

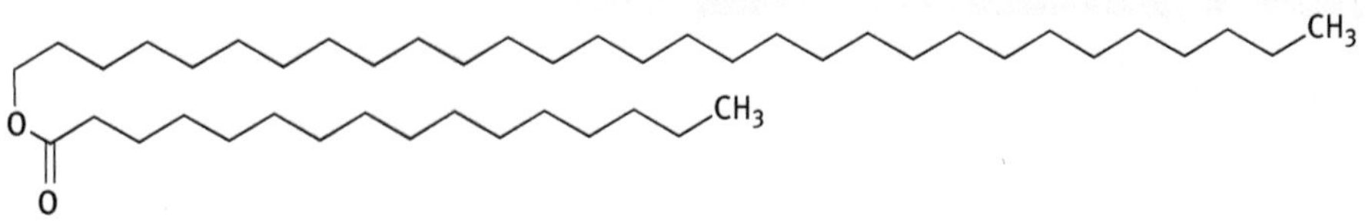

Figure 10.3: Lewis structure of beeswax.

10.5.2 Uses of beeswax

The general public tends to consider beeswax as a source for candle wax, but there are many other potential uses as well, some traditional, some much more modern. Many are small-scale uses, or are applications seen by end users in small amounts, such as wax for surfboards. A partial list includes the following:

1. Candles
2. Cheese casings
3. E901 food additive – it is generally recognized as safe (GRAS)
4. Furniture polishes
5. Personal care products – lip balms, lip glosses, eye liner
6. Surfboard coatings

Undoubtedly, there are other uses for beeswax as well.

While beeswax can be sold in large quantities for the large-scale production of an item like beeswax candles, it can also be sold in smaller amount, directly to consumers. Figure 10.4 illustrates samples of beeswax that can be purchased in health food stores.

The different colors of beeswax are not necessarily an indicator of its purity. Color and feel can be slightly different depending ultimately on the diet of the hive that produced it. As well, the wax of more than one hive can be mixed before it is packaged for sale.

Figure 10.4: Beeswax.

References

[1]   Cochineal Color Peru. Website. (Accessed 24 April 2024, as: https://cochineal.org/).
[2]   The Indian Silks Export Promotion Council. Website. (Accessed 24 April 2024, as: https://www.thein diansilkexportpromotioncouncil.com/).
[3]   The Silk Association of Great Britain. Web. (Accessed 24 April 2024, as: http://www.silk.org.uk/).
[4]   UIA, International Silk Association. (Accessed 24 April 2024, as: https://uia.org/s/or/en/1100039254).
[5]   Chemical & Engineering News. Bolt Thread Joins with Patagonia, May 16, 2016, p. 16.
[6]   Michigan Beekeepers Association. Website. (Accessed 24 April 2024, as: https://www.michiganbees. org/).
[7]   Texas Beekeepers Association. Website. (Accessed 24 April 2024, as: http://texasbeekeepers.org/).
[8]   Northern Nevada Beekeepers Association. Website. (Accessed 24 April 2024, as: https://www.north ernnevadabeekeepersassociation.org/).
[9]   Washington State Beekeepers Association. Website. (Accessed 24 April 2024, as: https://www.wasba. org/).
[10] Nebraska Beekeepers Association. Website. (Accessed 24 April 2024, as: https://nebraskabeekeep ers.org/).
[11] Florida Beekeepers Association. Website. (Accessed 24 April 2024, as: https://www.flstatebeekeep ers.org/).
[12] Vermont Beekeepers Association. Website. (Accessed 24 April 2024, as: https://www.vermontbee keepers.org/).

11 Flavors

11.1 Introduction

The idea of mass-producing organic molecules that are either structurally the same as those that are natural sources of flavors, or that mimic the aroma and taste of some substance even though it is not the same molecule, is a relatively new field of endeavor for chemists. The short history the field has does not mean though that it is small. Numerous foods have some flavor specifically added to it, and several nonfood materials also utilize one or more flavors. For example, the very simple molecule benzaldehyde has become a major source of artificial almond flavoring, as well as the starting material for several other, more complex organic molecules. But production of this on a large scale does not require a biosynthetic pathway; rather, toluene is used as a starting material, which means that ultimately this material comes from crude oil. Table 11.1 is a nonexhaustive list of flavor molecules and the flavor or flavors associated with them [1–3].

Some of the most widely produced flavors are discussed here. Each is made on a large enough scale that it is part of a greater industry of flavor chemistry.

11.2 Cinnamic aldehyde

Cinnamaldehyde was first isolated in the early nineteenth century, and as might be imagined, was derived from a natural product, cinnamon oil, for well over a century. It can still be manufactured from traditional organic synthetic steps, since its structure is not particularly complicated. Figure 11.1 shows the Lewis structure of cinnamaldehyde.

Today, numerous companies produce cinnamaldehyde, and while end uses may be for small consumer items – such as packs of chewing gum – the sheer economy of scale for such items means that it is produced on a multi-ton scale annually.

There are several different pathways to cinnamaldehyde, but one biosynthetic pathway starts with phenylalanine and in three steps arrives at the product. The first step utilizes phenylalanine ammonia lyase (often abbreviated PAL), followed by transformation with 4-coumarate CoA ligase (abbreviated 4CL), then followed by cinnamoyl-CoA reductase (abbreviated CCR) to give the final product, with the exocyclic double bond in the correct, *trans*-configuration. Figure 11.2 shows the highly simplified reaction, listing only the starting material and final product:

Other methods exist whereby cinnamaldehyde can be formed, including those simple enough to be run in an undergraduate teaching laboratory. Such may involve a distillation of what is called oil of cinnamon bark, or may start with benzaldehyde. In other words, cinnamaldehyde does not require a biosynthetic pathway to be

https://doi.org/10.1515/9783111330259-011

Table 11.1: Artificial flavors.

Flavor/odor	Chemical	Structure
Almond/cherry	Benzaldehyde	
Apple	Manzanate, malic acid	
Banana	Isoamyl acetate	
Butter	Acetoin, acetylpropionyl, or diacetyl	
Cinnamon	Cinnamaldehyde	
Fruit	Ethyl propionate	
Grape	Methyl anthranilate	
Orange	Limonene	
Pear	Ethyl decadienoate	
Pineapple	Allyl hexanoate	

Table 11.1 (continued)

Flavor/odor	Chemical	Structure
Sour	Citric acid	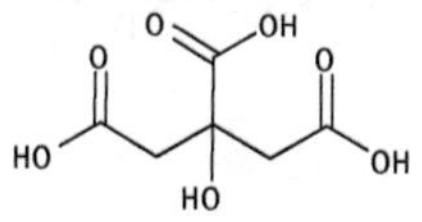
Spearmint	Carvone	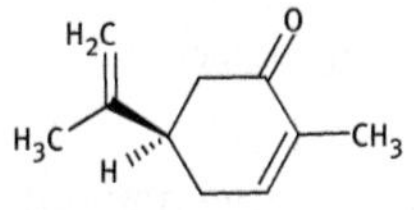 (R isomer)
Vanilla	Vanillin, ethyl vanillin	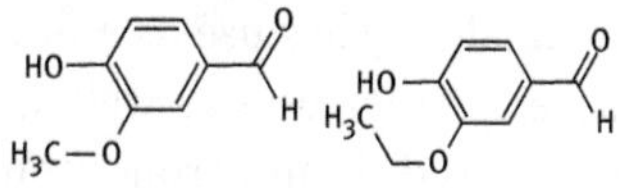
Vinegar	Acetic acid	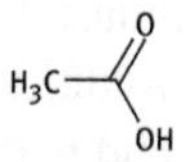
Wintergreen	Methyl salicylate	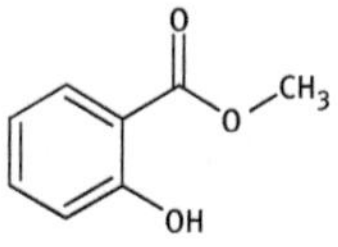

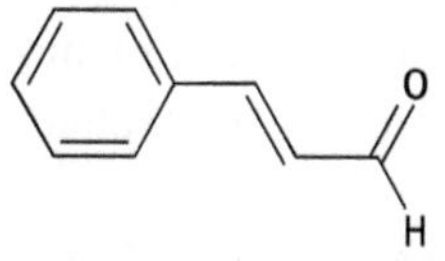

Figure 11.1: Lewis structure of cinnamaldehyde.

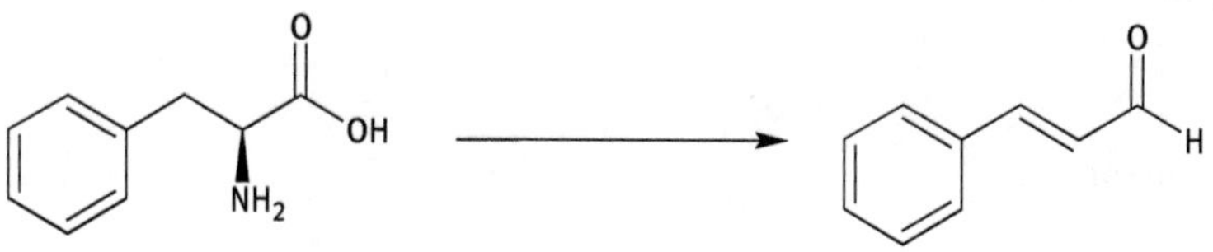

Figure 11.2: Cinnamaldehyde production.

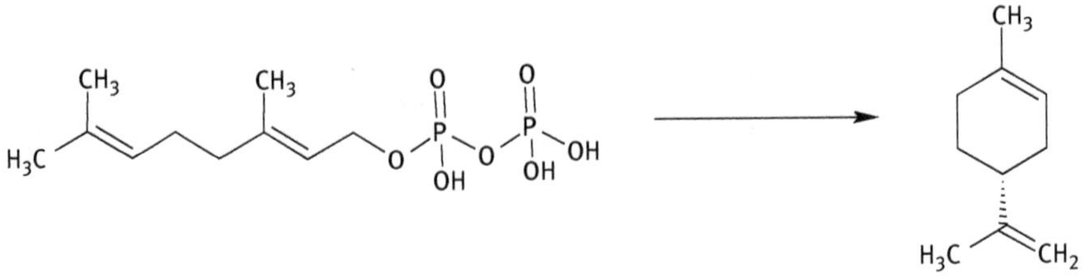

Figure 11.3: Limonene production.

formed in relatively large amounts. Indeed, the choice of production routes tends to remain one in which cost can be the deciding factor.

Cinnamaldehyde's use as a flavoring in gums and other foods is widespread, but it has several other uses as well. It has been found to have antimicrobial properties, and thus is used in numerous mouthwashes. It can be blended with other aromatic materials and used in perfumes. It has also found use in the agricultural sector as a fungicide and insecticide. Additionally, it has been used to inhibit corrosion on steel surfaces while such objects and parts are subjected to corrosive environments.

11.3 Limonene

Limonene, shown in Figure 11.3, is another organic molecule that is not particularly large and can be produced in a variety of ways. It provides the flavor of oranges, and can be added to foods as well as other materials when the scent is needed. One bio-synthetic pathway for the production of limonene starts with geranyl pyrophosphate, although more than one pathway exists to produce this product.

Curiously though, limonene has another large scale use – as a degreaser. This is the reason for its older, less formal name: citrus stripper oil. A more complete list of uses for limonene includes:
- Orange flavor
- Degreaser
- Fungicide
- Carvone precursor

Because of the relative simplicity of the limonene molecule, and the fact that the purity level of degreasers can be significantly lower than that for a fragrance of flavorant, there is limited incentive to produce limonene exclusively through biosynthetic path-ways. A large portion of it is still produced via traditional organic synthesis.

Additionally, limonene is transformed into the terpenoid carvone on a large scale. Carvone has found use for decades as a flavoring for spearmint chewing gum, and in numerous other applications where a mint scent or flavor is required. Figure 11.4 shows the Lewis structure of carvone.

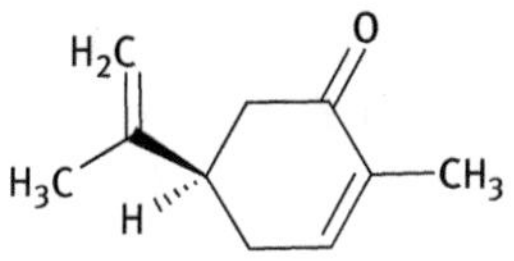

Figure 11.4: The structure of carvone (R isomer).

11.4 Vanillin

Vanillin, the flavor of vanilla, is probably the most widely used flavor in several markets. Traditionally, the substance has been extracted from vanilla beans, but the increasing demand for it, in foods, as well as other areas in which flavors are desired (such as candles and potpourri) means that some method of producing a synthetic vanillin is increasingly desired. Because the structure is relatively simple, at least when compared with other organic molecules, several different routes to synthetic vanillin have been attempted [3–5]. Figure 11.5 shows the Lewis structure of vanillin.

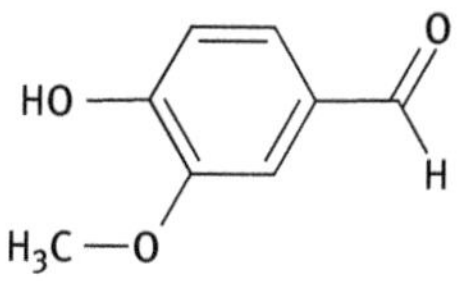

Figure 11.5: Structure of vanillin.

The organism *Vanilla planifolia* has been used in the recent past as a means of bypassing a traditional organic, multistep synthesis of vanillin. As with many biosynthetic pathways, this way of production minimizes the steps required in large-scale vanillin manufacture.

11.5 Reuse and recycling

All flavors, whether made by some biotechnological means or in some other way, are used in the production of some food or personal care product, and thus not recycled. The material that supports the flavorant is often a food, or some other consumable product, and thus there exists no means to recycle or reuse such materials.

References

[1]   International Food Safety Quality Network. Website. (Accessed 24 April 2024, as: www.ifsqn.com).
[2]   EWG. Website. (Accessed 24 April 2024, as: https://ewg.org/foodscores/content/natural-vs-artificial-flavors/).
[3]   Biotechnolgy in Flavor Production. Havkin-Frenkel, D. and Belanger, F.C., eds. Blackwell Publishing, Ltd., Oxford, 2008, 978-1-4051-5649-3.
[4]   Advanced Biotech. Website. (Accessed 24 April 2024, as: http://www.adv-bio.com/).
[5]   Mane. Website. (Accessed 24 April 2024, as: http://www.mane.com/biotechnology).

12 Plastics

12.1 Introduction

Perhaps obviously, the large-scale production of plastics has profoundly changed the way humans live in the world today and the quality of life we enjoy. The field has become so large that there are now several trade organizations devoted to the promotion of the use of plastics in almost every developed nation [1–13]. What can be called informally, "The Big 6" – those that are labeled with resin identification codes (RIC) 1 through 6 – are produced in such large amounts that the idea of replacing them with plastics produced in some greener way, or through some method utilizing biotechnological steps exclusively, may seem hard to envision. These six are as follows:

1. polyethylene terephthalate (PETE, or RIC 1),
2. high-density polyethylene (HDPE or RIC 2),
3. polyvinyl chloride (PVC or RIC 3),
4. low-density polyethylene (LDPE or RIC 4),
5. polypropylene (PP or RIC 5),
6. polystyrene (PS or RIC 6).

Figures 12.1–12.5 show the monomer units and the repeat polymer unit for these six plastics, respectively. Specific reaction conditions are not given, as there are usually several different sets of conditions, all of which produce slightly different version of the polymer, and some of which are proprietary to one company or another. The production of polyethylene can be undertaken at different densities, and it is difficult to represent the reaction chemistry of one exclusively. Rather, it is easier to state simply that LDPE has significantly more branches along its polymeric chains than does HDPE.

Figures 12.6–12.9 show several examples of how the RICs are placed on various consumer-end items. Figure 12.6 gives an example of PETE being used in a plastic cup or bottle, which the general public tends to think of as the primary use for PETE (as virtually all plastic beverage bottles are made from PETE). But it has several other uses as well, including a wide variety of various food packaging. Note that in Figure 12.7 the RIC code is at times located next to a trademark for a specific corporate name of a polyethylene. As well, note that in Figure 12.8, while polypropylene is not normally considered a material for use in foodware, cups can be made from it, as well as other end-user items. Additionally, in Figure 12.9 the RIC for polystyrene can be seen, in its use as Styrofoam®, although polystyrene does not have to have air blown through it to make it into usable consumer items.

The large-scale production of these six plastics almost always involves a series of chemical transformations starting with separations of small molecules from crude oil. A major challenge remains how to produce such materials using biobased starting materials.

https://doi.org/10.1515/9783111330259-012

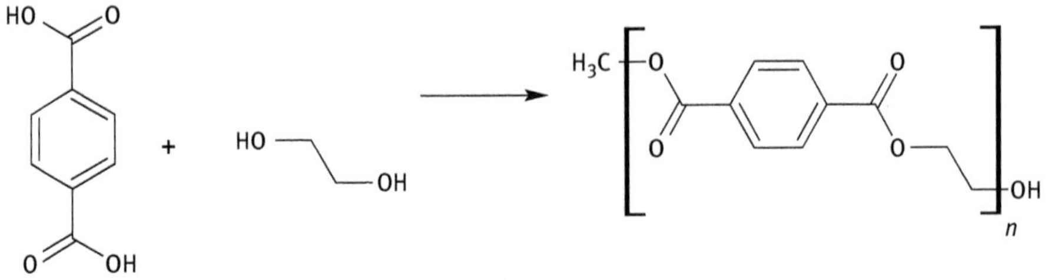

Figure 12.1: Synthesis of polyethylene terephthalate.

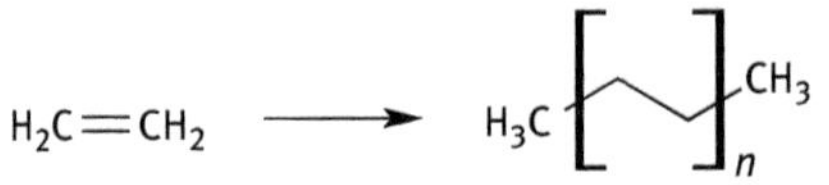

Figure 12.2: Production of polyethylene.

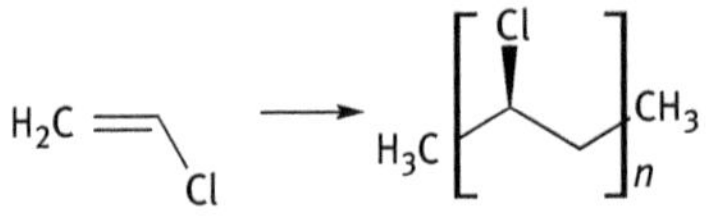

Figure 12.3: Production of polyvinyl chloride.

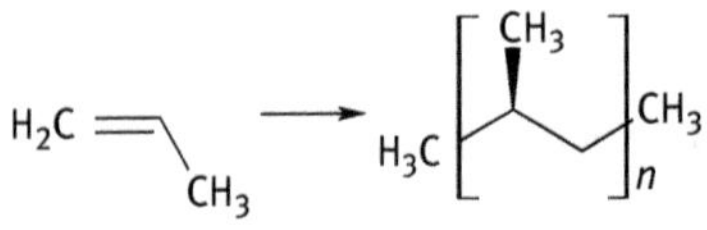

Figure 12.4: Production of polypropylene.

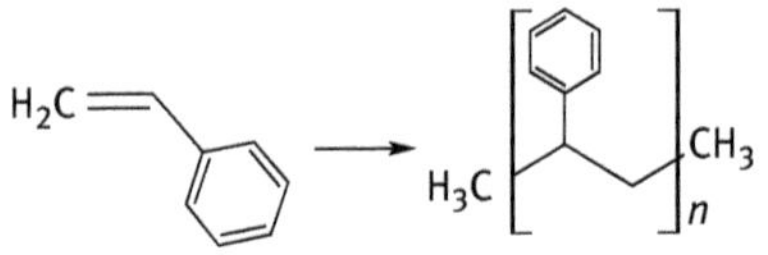

Figure 12.5: Production of polystyrene.

When these six plastics, as well as several others, were first produced on a large scale and marketed in a wide variety of end-user applications, they were sold to the public as materials that would last "forever." Undoubtedly, they are robust materials that do not degrade quickly. But the concept of a material or an item that lasts "forever" means that at least in theory, once enough of them have been sold, there is no need to produce more. This runs counter to the idea of continued production that underpins modern society.

To use a specific example, if companies make robust, reusable plates from a plastic, once enough plates have been made that everyone who will ever buy such plates has done so – and the plates are made to last "forever" – either the market for

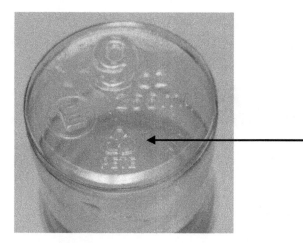

Figure 12.6: RIC 1, a polyethylene terephthalate cup.

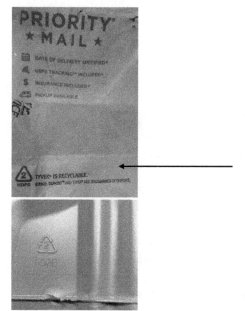

Figure 12.7: RIC 2 on a polyethylene United States Post Office envelope.

plates will cease to exist and the companies making them will have to close or shift to other products, or such plates will have to be marketed as disposable. While the latter option may seem to be a normal way to think now, when new products made from extremely durable plastics were first marketed, the market for them appeared to be limitless.

Figure 12.8: RIC 5, polypropylene.

Figure 12.9: RIC 6, on a polystyrene cup.

The above example is only one of thousands that have ultimately resulted in an enormous amount of plastic items and materials ending up either in landfills or in the world's oceans, so much so that the topic is now repeatedly the subject of articles in the popular press [14–17]. For this reason and others there has been a growing interest in the past decade to produce what are called biodegradable plastics, and to do so from renewable feedstocks, using microorganisms where possible in the production process. The four plastics we discuss here are four that have been aggressively developed and marketed as being in some way greener than traditional plastics, in part because of their use of some biotech step, often some microorganism used in a fermentation step.

12.2 Polylactic acid

Lactic acid, the precursor to polylactic acid (PLA) [18], is often sourced from corn, which provides sugars that can be converted to the starting monomer in a variety of ways, including via various microbes. The subsequent production of PLA from the lactic acid precursor usually involves metal catalysts, and depending on the conditions can be changed into an atactic or a syndiotactic form.

Figure 12.10 illustrates both lactic acid and polylactic acid, the basic repeat unit of the polymer.

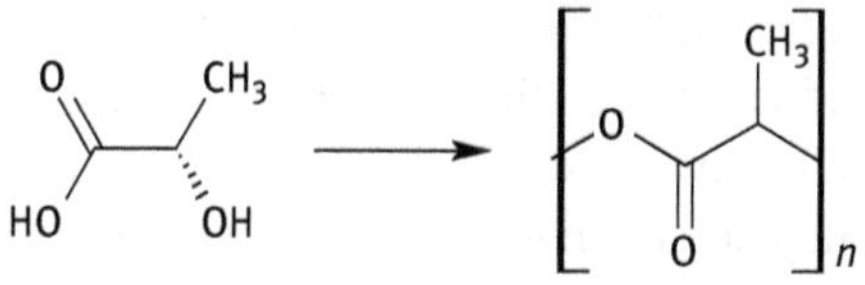

Figure 12.10: Lactic acid and polylactic acid.

One of the reasons for the rise in the use of PLA is that those marketing it can state truthfully that it is a biobased polymer that is not directly dependent upon petroleum. This has become an important point for consumers who want to avoid using plastics that come from crude oil, and that do not degrade. Figure 12.11 shows an example of user-end items made from PLA. Often items such as this disposable cutlery is marked with some indicator that it is not traditional plastic. In this case, the words "Eco Products®" appear on each item.

Figure 12.11: Plastic cutlery made from PLA.

While this important point about PLA coming from plant sources is true – the starting lactic acid is usually made from corn and not petroleum – there is a less direct but still vital connection to petroleum involved in this production. It is as follows:

- Modern corn varieties must be grown with fertilizer.
- Virtually all agricultural fertilizer used on an industrial scale is either ammonia (NH_3) or some derivative of it.
- In turn, ammonia is always produced by the Haber–Bosch process in which elemental hydrogen and elemental nitrogen are combined at high temperature and pressure, using an iron catalyst.
- The only economically viable method of producing elemental hydrogen on a large scale is from the gaseous fraction of crude oil production – methane, ethane, and propane.

Thus, the process called hydrocarbon stripping is used to make hydrogen that makes the ammonia that fertilizes the corn that produces the lactic acid. The end result is that PLA and other plastics sourced from corn or any agricultural products that must be fertilized are connected to petroleum and dependent on its extraction, but through what might be called several steps of separation.

12.3 Production of polyacrylonitrile

Polyacrylonitrile (PAN) [19] is another polymer with many industrial uses; and has recently seen more than 10 billion pounds produced annually. Virtually all of this is consumed in the production of plastics and various resins.

Recently, a single report has been published indicating that the starting acrylonitrile can be produced from 3-hydroxypropionic acid, which in turn is produced through a microbial pathway from glucose [20]. More established means of production that use some biotechnological step, usually a fermentation, include the use of lignocellulosic biomass, the use of the amino acid glutamic acid, and the use of glycerol. The latter has been an area of extensive focus, in large part because glycerol is a byproduct in the production of biodiesel, after the saponification of triglycerides, and is extremely inexpensive. All these compete against the well-established SOHIO process (from Standard Oil of Ohio), which produces acrylonitrile chemically, starting from propylene.

The production of PAN is an industrial polymerization process, which can proceed either via a free radical polymerization route or an anionic polymerization route. Figure 12.12 shows the repeat unit of PAN.

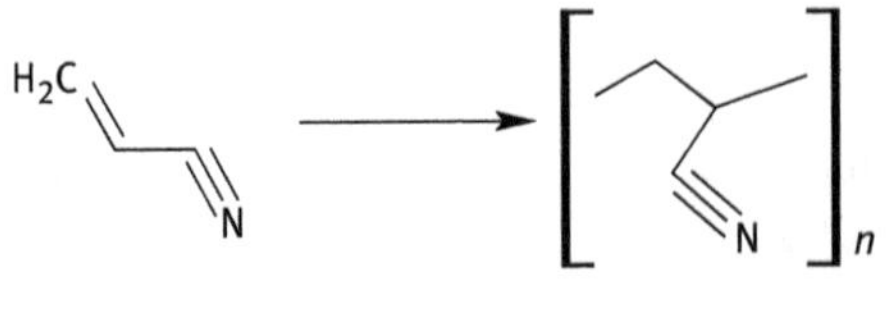

Figure 12.12: Acrylonitrile and PAN.

12.4 Polyhydroxyalkanoates

Polyhydroxyalkanoates (PHA) represent a class of polyesters that have been found to be made in nature by a variety of organisms. Bacterial fermentation can produce them, and because of this the source materials for PHAs are almost always sugars such as glucose or lipids. Because microorganisms can produce them, their separation from different cultures has been optimized in over two decades of experimentation. Since many different organisms produce them, there has been a continued effort to produce transgenic microbes or plants to optimize such yields [21]. When polymerized, these materials tend to be biodegradable, making them a preferred type of plastic in several applications [22]. Figure 12.13 shows a representative repeat structure of a PHA, although it should be emphasized that there are many different isomeric possibilities.

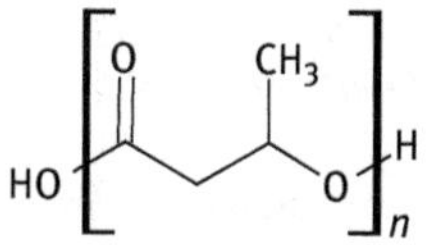

Figure 12.13: Structure of one of the polyhydroxyalkanoates.

12.5 Polyols

Polyols are not a single molecule, but rather a class of organic materials with multiple hydroxyl groups. In various polymer polymerization processes, polyols often act as cross-linkers. When reacted with isocyanates, they produce polyurethanes, a class of material with a large suite of uses, often as foams.

Cargill is a company that has been very successful in marketing polyols that have been produced from biological sources, specifically soybean oil, in what they have labelled their BiOH® polyols [23]. The company emphasizes the biobased sourcing of their polyols, and further emphasizes the savings this produces, when compared to the amount of crude oil that would otherwise be used in their production. But they do keep their source for creating the polyols proprietary.

Another company that has had success with the marketing of polyols is the Dow Chemical Company with their entire Renuva®™ line of polyols [24]. Much like their competitors, they advertise that their polyols are from a "natural oil-based" material, meaning some plant or animal source. But here as well, the process by which Renuva® is made remains proprietary.

In these cases, as well as the many which involve polyols made from petrochemical sources, the major use for them is the production of the just-mentioned polyurethanes, and the foams that are used in cushions, furniture, pillows, carpeting, and many other foam or padding applications.

12.6 Recycling and reuse possibilities

The production of plastics from biobased sources represents a new way to produce one or more commodity chemicals, one that frees the producer from a petroleum-based feedstock. But the fact that the end product can be what we might call an established plastic means that the need for recycling remains the same as for any plastic made from a petroleum-based feedstock. Since the end plastic is molecularly the same, whether produced from biosources or petroleum sources, each should be equally nonbiodegradable. Some materials, such as PHAs, avoid this, as they are themselves biodegradable.

References

[1]   United States Plastic Corporation. Website. (Accessed 24 April 2024, as: https://www.usplastic.com/).
[2]   Plastics Industry Association. Website. (Accessed 24 April 2024, as: https://www.plasticsindustry.org/).
[3]   Chemistry Industry Association of Canada. Website. (Accessed 24 April 2024, as: https://www.plastics.ca/).
[4]   British Plastics Federation. Website. (Accessed 24 April 2024, as: https://www.bpf.co.uk/).
[5]   Plastics Europe. Website. (Accessed 24 April 2024, as: https://www.plasticseurope.org/en).
[6]   Germany Trade and Invest. Website. (Accessed 24 April 2024, as: gtai.de/en/invest/industries/industrial-production/plastics-industry-68692).
[7]   Singapore Plastic Industry Association. Website. (Accessed 24 April 2024, as: https://www.spia.org.sg/).
[8]   Plastic Industry Manufacturers of Australia. Website. (Accessed 24 April 2024, as: https://www.pima-org.net).
[9]   Plastics South Africa. Website. (Accessed 24 April 2024, as: http://www.plasticsinfo.co.za/).
[10] The Association of Plastics Recyclers. Website. (Accessed 24 April 2024, as: https://plasticsrecycling.org/).
[11] Plastics Recycling World. Website. (Accessed 24 April 2024, as: https://plasticsrecyclingworldexpo.com).
[12] Plastics Recyclers Europe. Website. (Accessed 24 April 2024, as: https://www.plasticsrecyclers.eu/).
[13] Plastic Machinery Manufacturers Association of India. Website. (Accessed 24 April 2024, as: https://www.pmmai.org/).
[14] The Ocean: The ocean is the origin and the engine of all life on this planet – and it is under threat. Website. (Accessed 24 April 2024, as: thegef.org/sites/default/files/publications/CI_Ocean_Factsheet.pdf).
[15] Plastic Soup Foundation. What is plastic soup? Website. (Accessed 24 April 2024, as: https://www.plasticsoupfoundation.org/en/plastic-problem/plastic-soup/).
[16] National Geographic: Forever is a long time. Website. (Accessed 24 April 2024, as: https://www.nationalgeographic.com/magazines/l/plastic/index-ps.html).
[17] Fauna & Flora: Saving Nature Together. Website. (Accessed 24 April 2024, as: https://www.fauna-flora.org/explained/how-does-plastic-pollution-affect-marine-life).
[18] Nature Works. Website. (Accessed 24 April 2024, as: https://www.natureworksllc.com/).

[19] Biron, M. Industrial Applications of Renewable Plastics: Environmental, Technological, and Economic Advances, Elsevier, ISBN: 978-0323480659.

[20] Karp, E.M., Eaton, T.R., Sanchez I Nogué, V., Vorotnikov, V., Biddy, M.J., Tan, E.C.D., Brandner, D.G., Cywar, R.M., Liu, R., Manker, L.P., Michener, W.E., Gilhespy, M., Skoufa, Z., Bratis, A.D., and Beckham, G.T. Renewable acrylonitrile production. Science, 2017, 358, 1307–1310.

[21] Snell, K.D. and Peoples, O.P. Polyhydroxyalkanoate polymers and their production in transgenic plants. Metabolic Engineering, 2002, 4, 29–40.

[22] Yield 10 Bioscience. Website. (Accessed 24 April 2024, as: https://www.yield10bio.com/).

[23] Cargill. Website. (Accessed 24 April 2024, as: https://www.cargill.com/bioindustrial/what-are-bioh-polyols).

[24] Dow. Website. (Accessed 24 April 2024, as: https://www.corporate.dow.com/en-us/science-and-sustainability/2025-goals/renuva-program.html).

13 Wood

13.1 Introduction

In the past few decades, biotechnology has expanded into the forestry industry, especially into the adaptation of specific types of trees to grow in less hospitable environments than what is their norm. The established way wood is produced and processed involves growing rows of trees in what are often called plantations, then harvesting them in a controlled manner, so that the growth cycle always matches the rate of harvest. For example, pine plantations are planted, nurtured, and fertilized when young, their limbs pruned at least annually as they mature, then a fraction harvested, that fraction matching the number of trees planted to match the harvest. From there, trees are milled into lumber, or processed into paper.

13.2 Wood production

There has been considerable discussion and concern in the past few decades about the cutting of old growth trees in the United States and Canada, meaning those that have never been logged. Because of concerns over the loss of old growth forest sections, as well as because of the cost required for extracting such trees, many types of wood and the products ultimately derived from wood are now produced in managed forests, the just-mentioned plantations [1, 2].

Biotechnology as applied to the growth and production of wood is often related to producing trees that contain less lignin. Lignin is a complex organic material that is difficult to break down. What can be called traditional chemical treatment of lignin does break it down, in many cases for further use in the pulp and paper industry, but produces significant amounts of long-term pollutants as well.

13.2.1 Means of genetic engineering

The traditional means of creating trees with traits desired by humans involves the creation and growth of hybrids. It is difficult to determine with certainty when this began, but it may have gone on for thousands of years. But the definition of biotechnology when applied to trees now means much more than that [3]. The following discussion divides tree and forest management into several broad categories.

https://doi.org/10.1515/9783111330259-013

13.2.2 Cuttings

Growing trees by using cuttings of a single tree is sometimes called clonal farming. The perhaps obvious advantage to this is that any tree with a specific desirable trait will be reproduced exactly, as many times as needed. A less immediate disadvantage is that the entire planted area is more susceptible to disease and parasites than any farmed area with multiple specimens, or with multiple species.

A fascinating example of how disease can invade and destroy what is called monoculture farming – in this case, rubber tree farming – is the failed attempt by the Ford Motor Company in the first part of the twentieth century to farm rubber trees to produce the rubber needed to make tires, from 1928 to 1934. The attempt was made in a portion of the Amazon in Brazil, where Ford had purchased a land area as large as some of the smaller states within the United States. Trying to replace jungle with a single species ultimately did not work or yield an amount of latex exudate from the trees that was profitable; and in 1945 the entire area was sold back to the Brazilian government [4]. Rubber will be discussed in more detail in Chapter 15.

13.2.3 Gene insertion

As has been discussed in previous chapters, genes can be introduced into seeds and seedlings to produce trees with desired traits. What is called cis-genesis involves implanting a gene into one species of tree that can be considered a cross-able gene. This means it may be possible to cross the gene from another plant species through a natural method involving no biotechnological steps. But genes from nonplant species can be introduced as well, using the insertion techniques mentioned in Chapters 7 and 9. Whatever gene is introduced, however, the goal is very often the production of mature trees with higher percentages of cellulose, and lower percentages of lignin.

13.2.4 Control of engineered trees

The biotechnological techniques that are shared by the pharmaceuticals industry and the forestry industry is generally not of concern to the public when thought of in terms of producing medicines in enclosed laboratories, but does become a matter of concern when trees are allowed to grow in the open. For convenience, we can divide the methods used as follows:
–   Lab only
    Trees that are genetically altered in some way are grown and cultivated in a strict laboratory setting. The chance of any novel genetic material spreading to a wider environment is essentially nonexistent.

- Controlled, overseen field experiments
  As the name implies, this is the small-scale planting of trees in the open that have undergone some form of genetic modification. But this is generally not a large-scale operation on an industrial scale.
- Large-scale planting, with monitoring
- Large-scale planting, without monitoring

These final two approaches to growing trees are those that arouse concern and at times animosity from the general public, because genetically modified plants have now been introduced to the open, in some cases without close monitoring. To be fair, such plantings are generally undertaken only with modified trees that have already been studied thoroughly, so that the risk of any cross-contamination is believed to be minimal. But perceptions and past experiences, as well as past problems, have eroded confidence in some segments of the public.

13.3 Pulp and paper production

The genetic modification of trees designed for wood pulp production and ultimately of paper often involves any change that decreases the amount of lignin produced and increases the amount of cellulose in the starting material.

The current production of wood pulp, a mature established industry, utilizes what is called the Kraft process (from the German word "Kraft," meaning strong). The process has slight variations depending on the starting wood, but generally involves the following:

1. Impregnating wood fibers. Wood chips must be fed into a digester with water, in a ratio of roughly 4:1, water to wood.
2. Pressure digestion. At elevated temperature (roughly 175 °C) and pressure (7–9 atm.), for 2–5 h, wood chips are cooked with what is called white liquor, a mixture containing NaOH, NaSH, and Na_2S.
3. Liquor recovery. This step is actually the separation of the usable pulp (for paper or cardboard) from the remaining material, which is now called either black liquor or brown liquor.
4. Blowing as well as volatiles collection. This step reuses the liquor by concentrating it and capturing the volatiles. This maximizes the recyclability of the liquid materials.
5. Screening. Material that is large enough that it is not in solution is next sieved out.
6. Cellulose washing. Multiple washes separate the pulp from the liquor, resulting in pulps of various grades.
7. Bleaching. This final broad step ensures lignin has been separated from the remaining material.

The design of the process is to separate and extract the lignin from all other materials, concentrating the pulp that can be used for the production of high grades of paper. But this list of steps does not particularly emphasize the chemicals used in the process. Table 13.1 is an attempt to show what is used. Perhaps obviously, the bioengineering of trees from either seeds or seedlings can be a potent means by which the volumes of these chemicals can be reduced, especially if the trees contain less lignin to start.

Table 13.1: Kraft process chemicals.

Chemical	Processing step	Comment
Water	Steaming heat of wood chips	
White liquor	Break down of lignin, digestion	
Black liquor	Nonpulp portion, contains organics	Routinely recovered, separated, reused
Calcium carbonate	For CaO production	Used for $Ca(OH)_2$ production, used for Na_2S production.
Sodium sulfate	For Na_2S production	Needed for S^{2-} to degrade lignin.

Table 13.1 gives some indication of the strong bases that must be used in the Kraft process. Such material must be used responsibly, and ultimately costs companies when it must be disposed of. Several national or international organizations exist that promote the responsible husbanding and growth of trees and forests [5–9]. Some of these are also organizations concerned with the industrial-scale production of paper [10–13].

13.4 Recycling and reuse

The recycling of paper is a major, established industry throughout the world, yet its normal operation does not require any biotechnological steps. However, the use of recycled paper products has become a point of pride for numerous industries. Figure 13.1 shows an example of a paper product – in this case a greeting card – made from recycled paper. Note that it advocates not only that the product is made from recycled materials, but that the source is also part of a sustainability initiative. Both are directed at consumer desires to purchase products that are environmentally friendly.

Likewise, paper recycling is now emphasized in numerous public, commercial, and educational environments. Figure 13.2 shows a set of recycling containers where

Figure 13.1: Recycled paper consumer product.

Figure 13.2: Paper recycling collection point.

a single trash can once would have been, with a specific bin for paper that can be recycled, as well as bins for plastic recycling. In both of these cases, there is also an economic driver for recycling papers. The recycled product is less expensive to bring to market a second time than virgin papers.

Additionally, the reuse of lumber has become an established part of a larger industry; but again, the use of some biotechnological step or process is not often required. Lumber scraps are often mechanically chipped to smaller pieces, then pressed with adhesive to make various grades of what are called particle board.

References

[1]   Canadian Forestry Association. Website. (Accessed 24 April 2024, as: https://www.cfa-international.org/NGO%20directory/DFA-325.htm).

[2]   Forest Biotechnology and its Responsible Use: A biotech Tree Primer by the Institute of Forest Biotechnology, by Adam Costanza and Susan McCord, Raleigh, North Carolina.

[3]   Burdon, R. and Libby, W. Genetically Modified Forests From Stone Age to Modern Biotechnology. 2007, ISBN: 978-0-89030-068-8.

[4]   Fordlandia: The Rise and Fall of Henry Ford's Forgotten Jungle City, Greg Grandin, ISBN: 978-1-84831-154-1.

[5]   Resources for the Future. Website. (Accessed 24 April 2024, as: https://www.rff.org).

[6]   Resources. Website. (Accessed 24 April 2024, as: https://www.rff.org/resources.org/archives/the-worlds-forests).

[7]   European State Forestry Association, eustafor. Website. (Accessed 24 April 2024, as: https://www.eustafor.eu/).

[8]   American Forestry Foundation. Website. (Accessed 24 April 2024, as: https://www.forestfoundation.org).

[9]   Australian Forest Products Association. Website. (Accessed 24 April 2024, as: https://ausfpa.com.au/).

[10] American Forest and Paper Association. Website. (Accessed 24 April 2024, as: https://www.afandpa.org/).

[11] ASPI. Association of Suppliers to the Paper Industry. Website. (Accessed 24 April 2024, as: https://www.aspinet.org).

[12] Pulp and Paper Technical Association of Canada, PAPTAC. Website. (Accessed 24 April 2024, as: https://www.paptac.ca/).

[13] Confederation of European Paper Industries. Website. (Accessed 24 April 2024, as: https://www.cepi.org/).

14 Leather

14.1 Introduction

Leather, the skins of animals treated so that they will not rot and can be used in a variety of long-term applications, has a long history, and has been produced in almost all cultures. Animal hides have been tanned and worn by individuals as clothing and shoes, hardened and used as armor, made into dwellings such as yurts and teepees, made into small watercraft, and employed in a wide variety of other ways. The origin of the production of leather has been lost to history, but based on its widespread use, variations of the tanning process appears to have been discovered numerous times in different parts of the world. Even with the inroads made by synthetic fibers and materials into the uses we have just mentioned, leather remains a material of choice for many people in many areas. Thus, while some uses of leather fade with time, others become more pronounced, and some stay relatively the same. An example of the first is firewood carriers, also known as firewood slings – a large square of leather with handles that was widely used in colonial America, to wrap around cut or split wood. This was common when cloth was expensive and time consuming to make, and wood could pull or snag on such clothing, and when firewood was the common form of fuel for heating homes. Relatively few are made or needed today, although they have not completely disappeared from use. An example of the second might be luxury purses, backpacks, and other fashion accessories. It is not uncommon for high-end, designer purses today to cost over $2,000. In such cases, it can easily be argued that the cost is not the leather or the treatment of the leather that makes the purse, but rather the name of the luxury brand attached to it. An example of the third is something simple, such as boots. Work boots and combat boots have been made for centuries, and now exist in a wide range of quality and prices. Examples of the second and third are shown in Figures 14.1 and 14.2.

Leather therefore continues to be a large enough manufacturing sector that applications exist in a very wide array. As well, there are numerous national or international trade associations devoted to its promotion and use [1–9].

The application of biotechnology to leather can also be considered part of the wider spread of biotech into a number of industries. While many of these producers are not considered biotech companies, their use of biotechnological techniques and practices is driven by the desire to increase profit while at the same time reducing inefficiencies or wastes in the overall process. Interestingly, in the case of leather production, the introduction of biotech may be welcome for a reason unrelated to any of the other fields we have thus far examined: traditional leather production stinks. We will discuss traditional leather production and tanning in some detail later, and evidence of this simple statement will be made abundantly clear there [10–14].

https://doi.org/10.1515/9783111330259-014

Figure 14.1: High-end leather goods store.

Figure 14.2: Example of common leather boots.

14.2 Traditional leather tanning operations

The production and tanning of leather is a labor-intensive and water-intensive process. No matter what the source is – what animal hide – there are several common steps that begin with removal of the hide from the animal carcass, and end with a finished piece of leather. Through most of history, this has been what can be called a cottage industry, with all the steps performed by only a few individuals. Today, the large-scale production of meat for human and animal consumption can be intimately connected with leather production, although in some cases, animal skins are the highest value commodity. A list of the traditional steps in leather production includes the following:

1. Hide curing – normally the hide is treated with a brine solution. This is sometimes referred to as salt curing.
2. Hide soaking – a step to sterilize the hide, as antimicrobial compounds and surfactants are used.

3. Liming – this step further swells the hide and denatures proteins. Its name is derived from the traditional use of calcium hydroxide in water to make the solution often called milk of lime.
4. Hair removal, dehairing, or unhairing – as the name implies, this step is designed to remove hair from the hide. Sodium sulfide has traditionally been used in this step.
5. Bating – this step involves further hair and material removal, sometimes called scudding. It also makes the hide pliable. This step has in the past been the most putrid, as it often involved rubbing the hide with some animal manure. This has in the past been called the "dung water" step in leather processing.
6. Degreasing – fats and grease are removed in this step, making the leather more supple and pliable. Various detergents and solvents have been used in this step in traditional operations.
7. Tanning – This step has used vegetable tannins in the past. Somewhat more recently, this step involves chromium (III) compounds such as chromium (III) sulfate hexahydrate.
8. Processing of wastes – as mentioned, these steps and indeed the entire process is water intensive, and the liquid wastes must be disposed of or neutralized in some way. The problem of proper disposal is exacerbated in nations with weak environmental protection laws, and in small tanneries.

It is to be emphasized that these are general steps, and that some leather sources may require deviations from what has been listed [11].

14.3 Biotechnological developments

Since the production and tanning of leather involves hides or skins, and thus extended protein networks, it should seem logical that enzymes may be substituted for the more traditional chemicals used in almost every step. Thus, Table 14.1 shows both the traditional and a more modern example of each step in the process of producing leather.

Because the production of leather is such a widespread industry, with operations in almost all nations, a large number of companies exist that sell one or more enzymes or combinations of enzymes that can be used in leather production [14]. Recently, there has been a rise in companies based in China and India that manufacture such enzyme products. Many companies keep the composition of their enzyme mixtures proprietary, yet some information is available at different corporate websites. For example, a single company in India markets its enzyme mixtures using a series of trademarked names, most including the prefix "palk-", included in Table 14.1. [12].

Certainly, the use of a variety of enzymes has taken some of the traditional stink out of the production of leather, and has made the industrial production of leather

Table 14.1: Leather production, comparison.

Process	Traditional method	Enzymatic alternative	Examples, comments
Hide curing	Brine solution		Requires salt water baths
Hide soaking	Surfactants	Proteases, carbohydrases	*Aspergillus parasiticus*. Enzymes decrease production time.
Liming	$Ca(OH)_2$	Proteases	SEBate Acid™. Reduces production time.
Hair removal	Na_2S	Proteases	Palkodehair™. *Aspergillus flavus*. Enzymes are ecologically safer.
Bating	"Dung-water"	Proteases	Palkobate™, Palkocid™
Degreasing	Detergents, solvents	Lipases	Ecologically better
Tanning	Vegetable tannin, $Cr_2(SO_4)_3 \cdot 12H_2O$	Vegetable sources	Decreases heavy metal pollution

Source: [11].

more hygienic. As with many of the other processes we have examined in this text, however, the choice to shift from traditional production methods to those heavily involved with some newer, biotechnological processes is often one made based on economic factors. In short, companies tend to manufacture their products in the least expensive manner.

14.4 Recycling

The recycling of leather is a considerably smaller industry than that for the recycling of metal or paper, but leather can be recycled. Often, used items or scraps are gathered, mechanically shredded into very small pieces, combined with other materials, such as rubber, as well as binders, and recast into some usable material. This prevents leather items from ending in landfills [15].

References

[1]   Leather and Hide Council of America. Website. (Accessed 24 April 2024, as: https://www.us leather.org).
[2]   United States Hide, Skin, and Leather Association. Website. (Accessed 24 April 2024, as: https:// www.leatherpanel.org/).
[3]   American Leather Manufacturing Association, Inc. Website. (Accessed 24 April 2024, as: https:// www.americanleather.com).

[4]   Euroleather. Website. (Accessed 24 April 2024, as: http://euroleather.com/).

[5]   UK Leather. Website. (Accessed 24 April 2024, as: https://leatheruk.org/).

[6]   Indian Speciality Chemical Manufacturers Association. Website. (Accessed 24 April 2024, as: https://iscma.in).

[7]   Australian Hide Skin & Leather Exporters Association, Inc. Website. (Accessed 24 April 2024, as: http://ahslea.com.au/).

[8]   KIAA Kangaroo Industry. Website. (Accessed 24 April 2024, as: http://www.kangarooindustry.com/products/).

[9]   European Leather Association COTANCE. Website. (Accessed 24 April 2024, as: https://www.cotance.com).

[10]   The Leather Connection. Website. (Accessed 24 April 2024, as: https://www.leather.com.na).

[11]   Leather Industry Enzymes. Website. (Accessed 24 April 2024, as: https://leatherenzymes.weebly.com/degreasing.html).

[12]   Maps Enzymes Limited. Website. (Accessed 24 April 2024, as: http://www.mapsenzymes.com/).

[13]   Creative Enzymes. Website. (Accessed 24 April 2024, as: https://www.creative-enzymes.com/about.html).

[14]   Leather International. Website. (Accessed 24 April 2024, as: http://www.leathermag.com/).

[15]   BusinessRecycling.com.au. Website. (Accessed 24 April 2024, as: https://businessrecycling.com.au/recycle/leather).

15 Types of rubber

15.1 Introduction

Rubber is a material that has been known in some societies since ancient times (such as those in Central and South America), but has suffered for most of that time from being brittle when it becomes cold, and runny and difficult to form when it is hot. It is a polymeric material made from repeating isoprene units (more formally, 2-methyl-1,3-butadiene). A simplified example of both isoprene and the/ polymeric repeating units is shown in Figure 15.1.

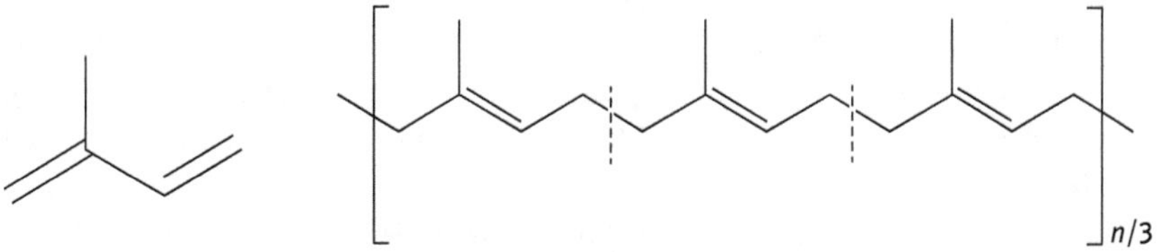

Rubber. The dashed lines divide the three isoprene units shown, which is why n is divided by 3.

Figure 15.1: Repeat unit of rubber.

Note that even as a polymerized material, there is a significant amount of unsaturation in rubber. This is noteworthy because the process of rubber vulcanization – the cross-linking of chains with sulfide bridges – requires sites where such links can form.

The production of rubber today has become an enormous, international business, and several organizations exist to promote the wide variety of uses and applications of rubber [1–3]. It is fair to say the public probably thinks of tires as the main use of the material, but there are numerous others as well. A nonexhaustive list would include:
- Clothing, boots, gloves, rain wear
- Conveyor belts
- Gaskets
- Hoses
- Mountings (for machinery with moving parts)
- Tubing
- Windshield wipers

As mentioned, this list is hardly exhaustive and complete; and new uses for rubber are still coming to the market. It should also be noted that what has been listed here constitutes a combination of what is called natural rubber and synthetic rubber, which is discussed in more detail later.

https://doi.org/10.1515/9783111330259-015

15.2 Natural rubber

Virtually all of what can be called natural rubber today comes from the *Hevea brasiliensis* tree, originally from the Amazon basin in South America, with only small amounts ever coming from the many other plant species that can produce some amount of latex, such as several types of vine of the *Lanophia* genus, such as *L. kirkii*. This latter has sometimes gone by the traditional name Congo Rubber and found its greatest use predominantly in the early part of the twentieth century.

Although the *H. brasiliensis* species is originally from the Amazon basin, it has been exported to all parts of the world in which the climate is suitable for growing the trees, making nations such as India one of the world's largest rubber producers today. Indeed, what can be called the illegal export of rubber seeds from Brazil in the 1870s is the subject of the book, *The Thief at the End of the World*, which details the exploits of Henry Wickham, the man responsible for bringing rubber tree seeds to Kew Gardens in England, the springboard from which the species was then exported worldwide [4].

Rubber trees are grown for the latex they produce. This sticky substance is extracted from the tree by scoring or tapping it – allowing the liquid latex to ooze out – a process that does not kill the tree. Rather, the reverse is true – the careful extraction of latex from a rubber tree actually causes it to produce more latex. This is done numerous times, making the trees very profitable over the course of years.

As with other trees, discussed in Chapter 13, the genetic manipulation of rubber trees is to a specific end. In this case, the production of more latex is that end. Even before the first vulcanization of rubber in 1839, it was recognized that those trees which produced the most latex were economically the best, and thus the traditional slow manipulation and growth of the best trees was undertaken. It should be noted that the life cycle of a rubber tree is such that it does not usually produce latex until it is 7 years old, although it can produce latex for many years after this point.

In 1987, a US patent was filed claiming to isolate pure rubber polymerase as a potential means to produce rubber that was the equivalent to natural rubber [5]. Interestingly, the enzyme is itself isolated from the *H. brasiliensis* tree's latex.

15.3 Dandelion rubber

The cultivation and use of dandelions to produce the latex for rubber production has been newsworthy of late [9, 10], but actually has a history reaching back to the Second World War, when the Axis Powers were in control of much of the land that produced rubber using the *H. brasiliensis* trees. What is sometimes called the rubber root, or the Kazakh dandelion (more specifically *Taraxacum kok-saghyz*) was cultivated at that time in attempts to produce a natural rubber to which the Allied Powers had unlimited availability. The end of the war meant the end of this effort, since the cost of the more traditional natural rubber again decreased, as did the need for it by militaries throughout the world.

Recently, reports have again discussed the use of latex from the dandelion plants, with those in the popular press sometimes not being particularly specific about species and cultivated varieties, often called cultivars, [6–9] for the production of rubber from sources that can be considered domestic in a rather large number of countries.

Recently, as early as 2013, a cultivar was developed at the Fraunhofer Institute for Molecular Biology and Applied Ecology (often IME) which enables the production of dandelion latex, and thus rubber, at what is close to an economically competitive level [9, 10]. This required the inhibition of a specific enzyme, and has been developed to the point at which IME and Continental Tires jointly began a pilot plant. Continental now proudly advertises at its website that it has produced tires from dandelion rubber, and does point out that once again this is the Russian dandelion that was used in the past [10]. Whether or not this will come to be a marketable alternative to rubber from the *H. brasiliensis* tree has not yet been determined.

As might be imagined, while the use of dandelion rubber is advantageous in that transport costs are substantially reduced, the existing problem in efforts at the wide growth of any dandelion species is simply concentrating and gathering the latex from an enormous number of small plants, as opposed to extracting latex from larger ones, basically trees. On the other hand, the life cycle of the dandelion is much quicker than that of the rubber tree, with the former producing latex in a single growing season. Additionally, *T. kok-saghyz* requires far less water than rubber trees, and can grow outside what has been called the "rubber belt," the warm, moist area in which *H. brasiliensis* thrives. All these factors are considered against what can be called the existing infrastructure of the mature, large-scale industry that is the growth of rubber trees.

15.4 Synthetic rubber

What has been called synthetic rubber has for decades been polybutadiene, or styrene–butadiene co–polymers (SBR), which have some different properties from natural rubber at the macroscopic level. An example of one basic repeat structure is shown in Figure 15.2.

Now, the broad term "synthetic rubber" can be used to mean a number of different materials, some of them containing chlorine, because chlorine is part of the monomer from which the end material is made. What the synthetic rubbers have in common is that petroleum is their ultimate source. Thus, the production of these materials is not subject to biotechnological manipulations.

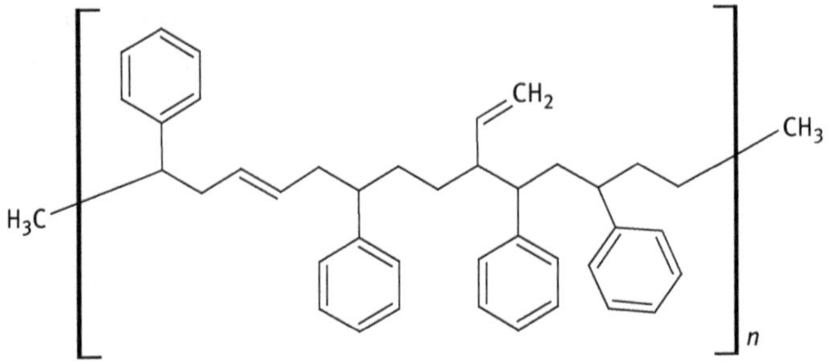

Figure 15.2: Example unit structure, SBR rubber.

15.5 Rubber recycling

Rubber recycling currently remains an area in which biotechnology has not yet made inroads on the industrial scale, even though it has been known for decades that rubber can be degraded with a variety of microorganisms. Rubber is often reused, but the term "recycle" when applied to other commodities generally means chemically or physically treated for some future use, which has proven difficult for rubber, since it is a series of cross-linked polyisoprene units, and thus an extremely large series of covalent bonds.

Similarly, the different synthetic rubber formulas are difficult to degrade chemically, and thus difficult to recycle. Note that in Figure 15.2 the entire structure is a series of covalent bonds, several of them aromatic, and all of them difficult to degrade.

Rubber reuse has become more common in roughly the past 20 years. Large rubber objects, very often tires, are shredded, and the shredded material is used in some manner. Examples that might be common to the general public are the use of shredded rubber as matting in children's playgrounds, as a component in athletic tracks, or as a component in rug matting [11]. Beyond this, even though it has been known for over 30 years that rubber can be degraded microbially, [12–14]. it appears that there are currently no large-scale rubber recycling programs.

References

[1]   U.S. Tire Manufacturers Association. Website. (Accessed 25 April 2024, as: https://www.ustires.org/).
[2]   Rubberworld Website. (Accessed 25 April 2024, as: https://rubberworld.com).
[3]   Association for Rubber Products Manufacturers. Website. (Accessed 25 April 2024, as: https://arp minc.com/).
[4]   Jackson, J. The Thief at the End of the World: Rubber, Power, and the Seeds of Empire. 2008, ISBN: 9780143114611.
[5]   Lui, J.J. and Shreve, D.S. US Patent 4,638,028. Rubber polymerases and methods for their production and use.

[6]   DuPont Life Sciences. Website. (Accessed 25 April 2024, as: https://www.dupont.com/).

[7]   International Service for the Acquisition of Agro-Biotech Applications. Website. (Accessed 25 April 2024, as: http://www.isaaa.org/).

[8]   Genencor. Website. (Accessed 25 April 2024, as: https://www.iff.com).

[9]   Turning Dandelions into Rubber: The Road to a Sustainable Future. Website. (Accessed 25 April 2024, as: https://www.scienceinschool.org).

[10] Continental. Website. (Accessed 25 April 2024, as: https://www.continentaltire.com/news/continental-constructing-tires-dandelions).

[11] U.S. Rubber Recycling, Inc. Website. (Accessed 25 April 2024, as: https://www.usrubber.com/).

[12] Tsuchii, A., Suzuki, T., and Takeda, K. Microbial Degradation of Natural Rubber Vulcanizates, Applied and Environmental Microbiology, 1985, 50(4), 965–970.

[13] Linos, A. and Steinbuchel, A. Microbial degradation of natural and synthetic rubbers by novel bacteria belonging to the genus Gordona. Kautschuk Gummi Kunstst, 1998, 51, 496–499.

[14] Linos, A., Reichelt, R., Keller, U., Steinbuchel, A. A Gram-negative bacterium, identified as Pseudomonas aeruginosa AL98, is a potent degrader of natural rubber and synthetic cis-1,4-polyisoprene. FEMS Microbiology Letters, October 1999, 182, 155–161. doi:10.1111/j.1574-6968.2000.tb08890.x.

16 Metals

16.1 Introduction

The extraction of metals from the earth is an ancient process, but also one that has seen numerous improvements in the most recent two centuries. Some metals exist in nature in what is called their reduced state. Gold, silver, copper, and iron are examples of this which have been known since ancient times. But virtually all of the elemental metals that have been discovered in the past 300 years have been found either alloyed with another element or elements, or have been found as some kind of ore, often an oxide or sulfide ore – or both. The techniques used to separate metals from such ores have become very well-established, mature processes, and none are considered dependent on any biotechnological steps. In the past few decades, however, there have been enormous inroads made into the metals industry whereby a biotechnological step is used in some way during metals beneficiation, extraction, or purification – often called bioleaching. Such steps generally represent an improvement in any process, since most metal refining chemistry involves producing stoichiometric or greater amounts of one or more unwanted byproducts, often a pollutant.

16.2 Traditional metal refining

A great deal of metal production is pyrometallurgical in nature, meaning it requires high temperatures to extract a metal from its ore. Copper, tin, and iron are three metals that have been so worked since ancient times. Indeed, the first two when alloyed gave their name to an age in several cultures – the Bronze Age – while iron gave its name to another, later one in most of those cultures – the Iron Age. But throughout most of history, the production of any metal was a small, intense cottage industry. The large-scale industrial metal refining we have become used to today is a result of the Industrial Revolution, and the scaling up of earlier known procedures.

While the enlargement and expansion of the metals industry – including the discovery of numerous metal elements in the 1800s – has changed human life today, it has come at a price: increased pollution, and some environmentally degraded, mined areas. The idea of using some bacteria or microbe at any step in the production of metals only can be traced back approximately 50 years, when it was recognized that organisms that can live in the extreme environments that exist in geysers and other such acidic areas might be useful in concentrating ores without the use of large amounts of chemicals that would later need to be recovered. Some of the first used were *Acidithiobacillus ferrooxidans*, *Acidothiobacillus thiooxidans*, and *Leptospirillum ferrooxidans*. All have since been extensively used in copper recovery.

https://doi.org/10.1515/9783111330259-016

16.2.1 Iron refining

Traditional metal refining begins with a metal ore, often a metal oxide or sulfide, and separates the metal in what is called a pyrometallurgical process, meaning an extremely high-temperature process, in which some reducing agent is added, often carbon, and in which naturally occurring impurities are refined out [1]. As just one example, Figure 16.1 shows the basic, simplified chemistry for the refining of iron from iron oxides. We say that the reaction chemistry shown is simplified because the iron ore never begins with all the iron in the same oxidation state, and there are often other elements present in small amounts, which are here considered impurities.

Reaction	Temperature range (°C)
$2 C(s) + O_2(g) \rightarrow 2 CO(g)$	200–700
$3 Fe_2O_3(s) + CO(g) \rightarrow CO_2(g) + 2 Fe_3O_4(s)$	600–700
$CO(g) + Fe_3O_4(s) \rightarrow 3 FeO(s) + CO_2(g)$	850–900
$CaCO_3(s) \rightarrow CaO(s) + CO_2(g)$	850–900
$FeO(s) + CO(g) \rightarrow Fe(l) + CO_2(g)$	1,000–1,200
$C + CO_2(g) \rightarrow 2 CO(g)$	1,300
$SiO_2(s) + CaO(s) \rightarrow CaSiO_3(l)$	

Figure 16.1: Iron production chemistry.

It can be seen from these reactions that a significant amount of carbon dioxide must be produced to produce liquid iron in its elemental state. Additionally, slag is produced in the refining process – $CaSiO_3$ and other silicates – all of which are waste products. When one considers the massive scale at which iron is produced and steel from that (roughly 80 million metric tons per year), as well as other metals, it becomes obvious how much byproduct production becomes a large-scale pollution problem.

Traditional iron refining does not involve any biotechnological component, simply because the just-outlined steps have been found to be the most economically feasible. As well, it is harder to utilize any organism that might react favorably with a metal when it is present as a metal oxide, versus a metal sulfide. Despite this, there have been inroads made in the past decade into what is called beneficiation of ores using some biological organism. This will be discussed in more detail below.

16.2.2 Copper and aluminum refining

Much like iron, copper and aluminum are refined from ores, with the main aluminum-bearing ore being bauxite, an aluminum oxide, usually with significant amounts of impurities (the red color of bauxite is due to iron impurities in the ore, for example). Copper exists in many ores, some of them sulfide ores instead of oxide ores. For both of these metals, an electrochemical process is used to purify the element to a point where it is industrially useful. In the case of copper, an acid bath is required, while in the case of aluminum – in what is called the Hall–Heroult process – the entire reaction is run in a molten state. Perhaps obviously, both processes are quite energy intensive.

Efforts at biotechnological improvements in copper refining in the past 50 years have focused on bacteria that can in some way consume sulfur in some oxidation state. This is because there are several copper sulfide sources throughout the world, and because these sources are often low in copper, and thus the use of bacteria can help concentrate the useful material in the ores. The biotech step then is an economic incentive toward greater copper recovery.

16.2.3 Scope of metals production

Metals production is perhaps obviously a sector of the mining industry (with oil production and coal production being other major sectors). It has become an enormous industry, especially as the human population has grown in the past century, and the standard and quality of life has increased. The size of multinational mining operations is enormous, encompassing an industry that is worth hundreds of billions of dollars, and that produces hundreds of millions of metric tons of refined metal each year. The decision to open a new mine is not one generally made by chemists or chemical engineers. Rather it is a corporate decision that usually involves a board, and that does have input from scientists and engineers employed in the company.

Some of the world's largest mining companies are listed in Table 16.1. It is noteworthy that their products vary widely, as do their locations worldwide. As well, it can be seen that some of these companies also have interests in energy-related materials, such as coal.

Much has been written about the recycling of metals, and while this industry has become large and matured significantly in the past four decades, recyclers tend to be small when compared to mining operations. Nevertheless, they exist because they both fill an economic need, and make a profit.

In both mining operations and recycling, keep in mind that the biotechnological steps we will discuss also are used because they make a profit for the company involved. If a process is cheaper to perform with a microbe or plant than with some purely chemical means, then that becomes the driving force to use such a method of metal recovery or extraction.

Table 16.1: Major mining companies.

Name	Interests	Headquarters	Comments	Website
BHP Billiton	Copper	Australia	Oil as well	https://www.bhp.com/
China Molybdenum	Molybdenum, tungsten, rare earths	China		http://www.chinamoly.com/en/
Glencore	Copper, cobalt, iron ore, nickel, zinc	Baar, Switzerland	Coal and oil as well	http://www.glencore.com/
KAZ Minerals	Copper	Kazakhstan		https://www.kazminerals.com/
Kumba Iron Ore	Iron ore	South Africa	Part of Anglo American Group	http://www.angloamericankumba.com/
Rio Tinto	Aluminum, copper, iron	Melbourne, Australia	Diamonds and energy also	http://www.riotinto.com/
Southern Copper	Copper	Mexico		http://www.southerncoppercorp.com/ENG/Pages/default.aspx
SQM	Lithium, potassium	Chile		http://www.sqm.com/
Sumitomo Metal Mining	Copper, gold, nickel	Japan		http://www.smm.co.jp/E/
Tianqi Lithium Industries	Lithium	China	Also lithium compounds	http://www.tianqilithium.com/en/about
Vale	Iron, nickel, manganese, copper	Brazil	Also coal	http://www.vale.com/EN/Pages/default.aspx
Vedanta	Aluminum, iron, copper, silver, lead, zinc	United Kingdom		http://www.vedantaresources.com/

Source: [2–14].

16.3 Biotechnological applications, biohydrometallurgy

Metals such as iron, as well as copper and aluminum, are found in large ore deposits and are refined into relatively low-cost commodities. But several other metals are not found in large deposits, are of higher value than these three, and must be in some way concentrated before it is economically feasible to use them. As well, in the past four decades, what are termed lower grade ores – those with smaller percentages of metals that do even include lower value metals such as copper – have been of interest for beneficiation and refining, simply because there is a steadily rising demand for them in end-user products. Examples of metals that are routinely concentrated during their extraction include gold, uranium, cobalt, and the platinum group metals.

Very broadly, what may be termed the life cycle or use cycle of metals can be divided into the following steps:
1. Beneficiation. – Metals or their ores are concentrated so they can be refined and purified.
2. Refining and reduction. – Ores are reduced with elemental metals as their end product. Some coproduct is always made as well.
3. Use. – Metals are used in some item, product, or process.
4. Distribution and disbursement. – Toward the end of their usable time, metals in some applications are dispersed back into the environment surrounding where they were used.
5. Recycling – The chemical and physical reshaping or rerefining of metals.

Hydrometallurgy is the use of aqueous solutions to extract metals from their ore, or from concentrates (soil and other solid mixtures containing a desired metal), without having to use high temperatures [14–24]. A significant amount of hydrometallurgy occurs without the need for some biological organism or some biotechnological step [11, 25]. A prominent example is the use of cyanide solutions to solvate gold from mining operations, which has been used since the mid-nineteenth century (none other than Carl Wilhelm Scheele first determined in 1783 that gold complexes with cyanide). Yet the past 40 years have seen significant advances in what is referred to as biohydrometallurgy, the use of biological organisms to recover and extract some economically useful metal or mineral from its starting state.

Use of some organisms such as bacteria or archaea to aid in this process finds its most profitable inroads in steps 1, 4, and 5 with some possibilities in step 2, in the lattermost case particularly sulfide ores. Widely used techniques include the use of microbes in soil batches to help concentrate a metal, or the use of plants to remediate soils into which metals have been dispersed – the former a process called bioleaching, and the latter a process called phytoremediation. A growing area in which microbes are now being used is the recovery of metals from e-wastes, such as the metal from circuit boards that have been discarded for recycling.

16.3.1 Bioleaching

What is called dump leaching and heap leaching involve the use of an aqueous solution on a pile of mined materials that are usually low in concentration of one or more desired metals, and allowing a solution to soak through it over the course of time (sometime months). This process has been well established as a means of extracting metals from a wide variety of materials. A now well-established example practiced on a large scale is the extraction of gold using aqueous cyanide solutions.

Bioleaching involves the use of a wide variety of microbes that can survive in sometimes harsh aqueous solutions, and that affix metal ions while they go through the leaching process [25–34]. Copper can be extracted from low-grade ores in this manner, at times by the use of a microorganism and Fe^{2+} ions, the latter of which are oxidized to Fe^{3+} during the process. While several companies indicate they employ bioleaching as a means of metal recovery from low-grade ores, they do not routinely advertise which microbes they favor [2–13]. This is most likely for two reasons: the desire to protect their process, and the possibility that from one extraction to another more than one type of microbe may be required, depending on the starting ore batch.

A company, Billeton, does explain its bioleaching in some detail, but without disclosing specific microorganisms. At its website it comments:

> As near-surface oxide orebodies are depleted, mines are having to process deeper sulfide ores that are lower in metal grade and metallurgically more complex. Modern technologies that are simple, cost effective and environmentally sound are required for their extraction. In many cases, bioleaching is the most suitable process. Mintek's Biotechnology Division is a leader in this rapidly-developing field, having been involved in the development of bioleaching technologies for over 25 years. The division offers a full range of services for the evaluation and commercial implementation of bioleaching processes for gold, uranium and base-metal projects [2].

It can be inferred that even though base-metal recovery is important, valuable metals such as gold and uranium are prime targets for enhanced recovery. With many less valuable materials, such as copper ores, the use of an organism, such as one of the *Thiobacillus* genus, functions by oxidizing the sulfur in the copper ore from the sulfide to a soluble sulfate. Figure 16.2 shows the simplified reaction chemistry.

$$CuS_{x(s)} + H_2O \rightarrow Cu^{2+}_{(aq)} + SO_4^{2-}_{(aq)}$$

Figure 16.2: Sulfur oxidation in copper production.

The goal behind using some bacterium in bioleaching is to avoid the large amounts of chemicals that would otherwise be required to solubilize the copper, preparing it for reduction to the metal.

The organisms listed in Table 16.2 have been proven to be effective in bioleaching operations and have been mentioned at commercial websites, but represent only a small fraction of a large number of species that may be able to do so, a total number which continues to grow as researchers continue to study microorganisms that are now called extremophiles [31–34].

Almost all bioleaching starts with the biological organism feeding off the mineral – a sulfide mineral – then, as mentioned, some oxidation of iron in the +2 state to the +3

Table 16.2: Bioleaching, example organisms.

Name	Description	Potential use
Acidophilic flagellate PR1		
Acidothiobacillus thiooxidans	Bacterium	Sulfur removal
Aspergillus niger	Fungus strain	Cu, Ag, Au, Sn, Al, Ni, Pb, Zn remediation
Cupriavidus metallidurans		Gold and heavy metal extraction
Leptospirillum ferrooxidans	Iron-oxidizing bacteria	Iron concentration
Penicillium simplicissimum	Fungus strain	Cu, Sn, Al, Ni, Pb, Zn remediation
Sulfobacillus acidophilus		Iron reduction
Thiobacillus		Sulfur oxidation

Source: [27–30].

state. But the critical step is that the microorganism must interact with the desired metal – must be able to use or immobilize it, so that it can later be retrieved. The search for new microorganisms to do this is termed bioprospecting. This has become a continued effort on the part of some companies, as well as of some academic researchers, since the potential involved in this area is extremely large. In general, though, the steps involved in concentrating ores can be illustrated by the flow diagram in Figure 16.3.

16.3.2 Phytoremediation

Numerous plants take up one or more minerals as they grow. This has become of interest in the past decade as a possible means whereby polluted soil can be remediated, and the metals in such soils removed. Termed phytoremediation, the process involves sowing some plant, allowing it to grow, and in the normal process of its growth to extract one or more metals from the soil in which it grows.

Phytoremediation is being used increasingly to clean soils in which some past pollutant has accumulated. An example might be that of a downwind area from some metal refinery upon which smokes and soot have settled, thus polluting the area with one or more metals that can be of value if they are reclaimed. Such plants are termed hyperaccumulators. Based on this description, it may seem obvious that phytoremediation can only clean such a soil to the depth to which the plant's roots extend. For this reason, this form of remediation is not yet used on a large, industrial scale, although there are several promising efforts that have been attempted and examined in the recent past.

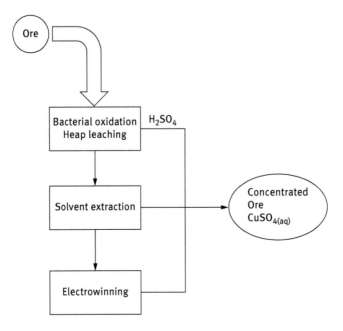

Figure 16.3: Concentration of low-grade ores.

References

[1]   Mining.com. Website. (Accessed 25 April 2024, as: https://www.mining.com).
[2]   BHP Billiton. Website. (Accessed 25 April 2024, as: https://www.bhp.com/).
[3]   CMOC Group, Ltd. Website. (Accessed 25 April 2024, as: https://en.cmoc.com).
[4]   Glencore. Website. (Accessed 25 April 2024, as: https://www.glencore.com/).
[5]   KAZ Minerals. Website. (Accessed 25 April 2024, as: https://www.kazminerals.com/).
[6]   Kumba Iron Ore. Website. (Accessed 25 April 2024, as: https://www.angloamericankumba.com/).
[7]   Rio Tinto. Website. (Accessed 25 April 2024, as: https://www.riotinto.com/).
[8]   Southern Copper. Website. (Accessed 25 April 2024, as: https://www.southerncoppercorp.com/).
[9]   SQM. Website. (Accessed 25 April 2024, as: https://www.sqm.com/).
[10] Sumitomo Metal Mining. Website. (Accessed 25 April 2024, as: https://www.smm.co.jp/).
[11] Tianqi Lithium. Website. (Accessed 25 April 2024, as: https://www.tianqilithium.com/).
[12] Vale. Website. (Accessed 25 April 2024, as: http://www.vale.com/).
[13] Vedanta Resources. Website. (Accessed 25 April 2024, as: https://www.vedantaresources.com/).
[14] Core Group. Website. (Accessed 25 April 2024, as: https://www.coreresources.com.au/).
[15] Natarajan, K.A. Biotechnology of Metals: Principles, Recovery Methods and Environmental Concerns. 2018, ISBN: 978-0128040225.
[16] Hocheng, H., Chakankar, M., and Jadhav, U. Biohydrometallurgical Recycling of Metals from Industrial Wastes, 2017, ISBN: 978-1138712614.
[17] Rubberdt, K., Glombitza, F., Sand, W., Schippers, A., Veliz, M.V., eds. 22nd International Biohydrometallurgy Symposium: Selected, Peer Reviewed Papers, Freiberg, Germany (Solid States Phenomena), DeGruyter, 2017.

[18] Abhilash Pandey, P.D. and Natarajan, K.A. Microbiology for Minerals, Metals, Materials and the Environment, 2015, ISBN: 978-1482257298.

[19] Schippers, A. and Sand, W. Biohydrometallurgy: From the Single Cell to the Environment (Advanced Materials Research), 2007.

[20] Teixeira, M.C. and de Carvalho, R.P. Biohydrometallurgy: Fundamentals, Technology and Sustainable Development, Part A (Process Metallurgy), 2001.

[21] Zhou, Q.G. Biohydrometallurgy: Biotechnology Key to Open the Door to the Use of Mineral Resources, 2000.

[22] Amils, R. and Ballester, A. Biohydrometallurgy and the Environment Towards the Mining of the 21st Century (Process Metallurgy), 1999.

[23] Rossi, G. Biohydrometallurgy, 1990, ISBN: 3-89028-781-6.

[24] Norris, P.R. and Kelly, D.P. Biohydrometallurgy, 1988.

[25] Dunbar, W.S. Biotechnology and the mine of tomorrow. Trends in Biotechnology, 2017, 35, 79–89.

[26] Minetek. Website. (Accessed 25 April 2024, as: https://www.mintek.com/projects/).

[27] Mining Technology. Toxic Mines: Benefits of Bioleaching Bacteria, Website. (Accessed 25 April 2024, as: https://www.mining-technology.com/features/feature122499/).

[28] The Invisible Miners. Website. (Accessed 25 April 2024, as: https://www.youtube.com/watch?v=6i_XfkCvsao).

[29] Mineral-Munching Microbes: The Future of Metal Mining? Website. (Accessed 25 April 2024: https://www.mining-technology.com/features/featuremineral-munching-microbes-futuremetal-mining/).

[30] Sitharashmi S., Kundu, M., and Behari sukla, L. Bio-beneficiation of iron ore using heterotrophic microorganisms, Journal of Microbiology and Biotechnology Research, 2015, 5(2), 54–60.

[31] Geller, W., Klapper, H., and Salomons, W., eds. Acidic Mining Lakes: Acid Mine Drainage, Limnology and Reclamation. Springer-Verlag, Berlin Heidelberg, 1998, ISBN: 978-3-642-71956-1.

[32] Copper Development Association Inc. Producing Copper Nature's Way: Bioleaching. Website. (Accessed 25 April 2024, as: https://www.copper.org/publications/newsletters/innovations/2004/05/producing_copper_natures_way_bioleaching.html).

[33] Vaughn, J., Riggio, J., Chen, J., Peng, C., Harris, H.H., and van der Ent, A. Characterisation and hydrometallurgical processing of nickel from tropical agromined bio-ore. Hydrometallurgy, 2017, 169, 346–355.

[34] Hurtado, C., Viedma, P., and Cotoras, D. Design of a bioprocess for metal and sulfate removal from acid mine drainage. Hydrometallurgy, 2018, 180, 72–77.

17 Textiles and detergents

17.1 Introduction

The production of textiles for clothing, sailcloth, and numerous other items has a history that stretches back into ancient times. Archaeological sites throughout the world have revealed a wealth of information about numerous peoples because of the textiles (as well as other items) found in burial sites. And while the production of textiles has been what can be called a cottage industry for most of history, the Industrial Revolution started at least in part a quest to produce large amounts of textiles for numerous uses.

17.2 Traditional production

There are only a few materials that have seen use in the production of textiles for centuries, and in some parts of the world for millennia. They are cotton, wool, silk, and leather. Silk was discussed in Chapter 10, when we examined insect sources of materials. Leather was the subject of Chapter 14. Cotton and wool have not been discussed thus far, and so a short introduction is necessary.

17.2.1 Cotton

The cotton plant is of the *Gossypium* genus, and the fluffy fiber that grows as what is called a boll which is then harvested is what the general public refers to as cotton. Cotton has been used extensively for clothing, blankets, and numerous other applications for thousands of years. For most of that time it was harvested and sorted by hand, with mechanization being brought to the process in the late nineteenth century. Even today, the production of cotton is a large enough industry that there are numerous organizations devoted in whole or in part to its production, use, and marketing [1–7].

What might be called the traditional improvements of cotton plants involves the slow process of selecting plants with favorable properties. In the past two decades, however, advances have been made in cotton production that are aimed not at large bolls on the plant, but at combatting insect pests, cotton bollworm, pink bollworm, and tobacco budworm [7]. The use of such varieties not only increases yield, and thus profits to the growers, but also decreases the use of artificial pesticides. This latter advantage is also an environmentally friendly improvement in the production cycle.

Despite the rise in the use of synthetic fibers, cotton remains a favorite for clothing, and in some other applications, because of the ease with which it can be spun and woven, and because garments made of it feel good against the skin.

https://doi.org/10.1515/9783111330259-017

The broad term "wool" means the hair of any animal, and indeed, throughout history, the hair of alpacas, llamas, goats, sheep, and several other species of animals have been used and spun as wool.

Much like cotton, the production and use of wool – generally meaning the wool of various types of sheep – is a large enough industry that there are several trade organizations devoted to it today [8–12]. Certain breeds of sheep are considered prized for the amount of wool they can produce. Wool has been used to make clothing in many cultures and many times because it is extremely warm, and can keep a person warm even when it is wet. As well, wool does not burn, which became extremely important as professional armies came into existence, and began using artillery. These new weapons shot residual fire as well as projectiles from the large guns. Men called cannoneers always wore wool pants and coats, to prevent stray sparks from igniting their clothing. Even today, although wool clothing can be considered heavy when compared to synthetic fibers, it remains a useful material and fabric in cold and other extreme environments.

17.3 Synthetic fibers

What are called synthetic fibers are generally those that have a petroleum base, such as the broad field of polyesters. The history of synthetic fibers is much shorter than those of materials such as cotton or wool, but especially since the Second World War has seen rapid growth. There are now several trade organizations related to the fashion industry, most of which deal in some way with the continually growing field of synthetic fibers and materials [13–20].

The field of synthetic fibers extends far beyond clothing and apparel, and includes almost all applications where some large woven, cast, or molded sheet is used. Examples include sails, tents, tarpaulins, and roofing materials. As well, the choice of materials has grown in the past few decades to include such as Gor-Tex® and other water-resistant materials, some of which can be considered fluorine-based as much as petroleum-based. Yet all of these are essentially plastics that have been spun into fibers, then woven into garments or made into other end-user items.

The industrial scale of synthetic fibers stretches back to a beginning in the 1930s, with the production of nylon. As synthetic fibers were brought to the market, they were touted as being stronger than traditional fibers, longer lasting, and more resistant to wear and to stains. Their one major disadvantage was their low melting point. But the low cost of their production, and their resistance to water and water damage (very important in the production of sails and other textiles used in a marine environment), made them extremely attractive to the general public.

17.4 Biotechnology and enzymes in textile production and detergents

A wide variety of enzymes effect some transformation that can have applications to textiles at some step of their production, processing or overall care. It is perhaps obvious as to why, since textiles can be made from cotton, wool, blends of these two fibers, or blends of these fibers and other types of material, and since materials spilled on them are often of the same types, meaning some matter that can interact with an enzyme, such as proteins, fats, and carbohydrates.

Traditional methods of cleaning stains and bleaching textiles involve a wide variety of chemicals, often in solution. Some of these are starkly nonnatural, such as the chlorinated dry cleaning material tetrachloroethylene (C_2Cl_4). The persistent nature of some of these compounds has caused long-term pollution problems in some communities, and thus the use of enzymes and other materials that can be deemed and called natural has become more common. Figure 17.1 shows the storefront of dry cleaners that promotes its business by appealing to this in its customers.

Figure 17.1: Dry cleaning storefront.

The smaller "eco" sign in the storefront states "100% natural product," indicating that what can now be called traditional dry cleaning fluids are not used there.

The following types of enzymes are discussed simply because they are already in use either in some production step for a textile, or in some care and cleaning step. Our discussion is somewhat general, since many companies choose not to release their formulas, especially of cleaning agents, and thus retain it as proprietary information. In the United States, since cleaners and detergents are not made for human consumption, there is no requirement to list ingredients on the packaging of such products.

17.4.1 Amylases

Amylases are a group of enzymes that effects the hydrolysis of starch to sugars, the monomeric units of starch. They are found widely in sources as different as mushrooms and various animal salivas.

In the textile industry, a variety of amylases have found use in what is called the desizing of a fabric. Often, starch is applied to some fiber before the fiber is woven into a fabric, to help strengthen it during what is often a high-speed process. This added material, this starch, needs to be removed from the cloth when the material is finished. Amylases are used for this removal, this desizing, because they selectively degrade the applied starch without degrading the resulting fabric.

As well, amylases can be used to remove stains, generally starch-based stains, from cloth. These can be used as a whitener, because removal of any stain generally improves the overall appearance of the fabric. Similarly to desizing, the amylase enzymes do this without degrading the fibers of the material [21].

17.4.2 Cellulase

Cellulase, a type of enzyme that is hydrolytic, breaks down cellulose into smaller molecules, and can ultimately degrade cellulosic fibers to glucose. It is found widely in animal gut, as well as in fungi and bacteria [22, 23].

The use of cellulases in a controlled manner effects breakdowns that can be felt – the textures of textiles are changed in the process. The use of cellulases has become widespread within the textile industry, and is now a significant percentage of all enzymes that are manufactured for a specific industrial use. To use an example, the use of cellulases is largely responsible for the feel of what are still called stone-washed jeans, even though stones are no longer used in the process [21, 22]. Cellulases are also widely used in stain removal, and are now found in several laundry detergents.

17.4.3 Lipase

Lipases are a class of enzymes that degrade fats. They are found in many animals, including humans. Lipases are also found in some microbes.

Lipases have a long, established use in the production of cheeses and yogurts. Much more recently, they have found use in detergents that have reached the market, once again for their ability to degrade and remove stains. They function by breaking the triglyceride-to-fatty acid bond.

17.4.4 Protease

Proteases are a class of enzymes that degrade proteins by hydrolyzing the peptide bonds within them. They are extremely widely spread as far as sources, including plants, animals, bacteria, and fungi.

As might be expected, certain proteases have been found to be extremely effective in stain removal. Since laundry detergents are formulated to remove what can be called common stains, that is, food stains (among others), they must be able to remove proteins as well as fats and carbohydrates, as discussed earlier. One protease that has found use in this area is subtilisin. The name is derived from *bacillus subtilis*, sometimes known as hay bacillus.

In detergent formulations, often several different enzymes are combined for the optimum effect. Table 17.1 shows a brief summary of the different types, and the target for which they are usually added to some detergent formulation. The field of proteases is particularly large, and has found wide use beyond that in detergents.

Table 17.1: Enzymes used in textile production and cleaning.

Enzyme class	Description	Use(s)
Amylase	α-Amylases – calcium metalloenzymes β-Amylases – maltose production γ-Amylase – glucose production Glycoside hydrolases	Starch to sugar degradation, also used in fermentations
Cellulase	Widely found in fungi, protozoa, and bacteria	Cellulose and polysaccharide degradation; can be used in biofuel production
Lipase	All, types of esterases	Fat and lipid degradation, also used in cheese production
Protease	Aspartic protease, cysteine protease, glutamic protease, metalloprotease, serine protease, threonine protease	Protein degradation and stain removal, also used in bread production

17.5 Polymers

Polymers used in textile manufacture are often those plastics made in large quantities, which have been covered in some detail in Chapter 12, and in Section 17.3. These have found extremely widespread use in clothing and other textile applications simply because they are inexpensive and easy to spin into fibers, and then weave into some textile. When made into clothing, these also acquire stains when in use, and can be treated with enzymatically active detergents.

References

[1]   National Council of Textile Organizations. Website. (Accessed 25 April 2024, as: http://www.ncto. org/).
[2]   American Fiber Manufacturers Association, Inc. Website. (Accessed 25 April 2024, as: https://www. omicsonline.org/societies/american-fiber-manufacturers-association-afma/).
[3]   Industrial Fabrics Association, International. Website. (Accessed 31 August 2018, as: https://www. ifai.com/. – and: https://www.ifai.com/?s=bio).
[4]   Canadian Textile Industry Association. Website. (Accessed 25 April 2024, as: https://www.canadatex tiles.ca/).
[5]   Cotton USA. Website. (Accessed 25 April 2024, as: https://cottonusa.org/).
[6]   International Cotton Association. Website. (Accessed 25 April 2024, as: http://www.ica-ltd.org/).
[7]   National Cotton Council. Website. (Accessed 25 April 2024, as: https://www.cotton.org).
[8]   American Sheep Industry Association. Website. (Accessed 25 April 2024, as: https://www.sheepusa. org/).
[9]   Canadian Cooperative Wool Growers Association. Website. (Accessed 25 April 2024, as: https://www. wool.ca/).
[10] European Wool Association. Website. (Accessed 25 April 2024 as: https://www.europeanwoolassocia tion.org).
[11] Australian Wool Growers Association. Website. (Accessed 25 April 2024, as: https://www.auswool growers.com.au/).
[12] New Zealand Wool Classers Association. Website. (Accessed 25 April 2024, as: https://woolclassers. org.nz/).
[13] Canadian Apparel Federation. Website. (Accessed 25 April 2024, as: https://www.apparel.ca/).
[14] Euratex, The European Apparel and Textile Confederation. Website. (Accessed 25 April 2024, as: https://www.euratex.eu/).
[15] ETP, Fibres Textiles Clothing, European Technology Platform. Website. (Accessed 25 April 2024, as: https://textile-platform.eu/).
[16] Industrieverband: Veredlung-Garne-Gewebe-Technische Textilien e.V. Website. (Accessed 25 April 2024, as: https://www.ivgt.de/en/).
[17] Textile Services Association. Website. (Accessed 25 April 2024, as: https://www.tsa-uk.org/).
[18] UK Fashion and Textile Association. Website. (Accessed 25 April 2024, as: https://www.ukft.org/).
[19] Specialised Textiles Association. Website. (Accessed 25 April 2024, as: http://specialisedtextiles.com. au/).

[20] Australian Textile Mills Pty., Ltd. Website. (Accessed 25 April 2024, as: https://www.australiantextile mills.com.au/).

[21] Monteiro de Souza, P. and de Olivera Magalhaes, P. Application of microbial alpha-amylase in industry – A Review, Brazilian Journal of Microbiology, 2010 Oct–Dec 41(4), 850–861.

[22] Polaina, J., MacCabe, A.P., eds. Industrial Enzymes: Structure, Function and Applications, Springer, Chapter 4: Cellulases in the Textile Industry, Arja, M-O., pp. 51–63.

[23] Van Beckhoven, R. F., Zenting, H. M., Maurer, K. H., Van Solingen, P., and Weiss, A. Bacillus cellulases and its application for detergents and textile treatment. European Patent. EP 739, 1995.

18 Cosmetic ingredients

18.1 Introduction

Cosmetics have been used for thousands of years to change the image of a human face, ultimately to what is perceived to be a more attractive version. Throughout this time, a wide variety of materials have been used to enhance eyes, lips, and skin – some of them rather toxic. Perhaps the most famous example of a toxic facial makeup is the lead-based white powders used by Queen Elizabeth I (and presumably others in her court, as well), which appear to have caused blood poisoning in her, although the cause of her death remains a debated topic.

More recently, a wide variety of organic chemicals have been used in the personal care products industry, in every aspect of the field, from formulations to solvents. Roughly since the end of the Second World War, most of these have oil as a source material, and come from the fractionation and distillation of oil. The industry has become so large that there are several trade organizations that advocate for it, in countries throughout the world [1–11].

In the last two decades, there has been a significant upswing in interest in having cosmetics made from some plant or animal source, as opposed to crude oil, thus making the material renewable, as opposed to fossil-fuel based. As well, different enzymes and biologically based materials have found use in widely different applications [12–14]. Arguably the most famous example is that of botox, a protein that can be injected under the skin to slow the progression of wrinkles. The bacteria *Clostridium botulinum* is the source of the protein. The botulinum toxin relaxes muscles underlying the skin, and thus smooths the skin.

The companies that are interested in some biotechnological aspect of personal care are constantly changing, and some are being bought by others. As well, the number of start-ups continues to grow, adding to the total and the diversity of offerings on the market. Table 18.1 shows a nonexhaustive list of such companies, as well as the products for which they are known.

It is shown in Table 18.1 that some of the firms listed are very specific in the production of a flagship product, while others produce a wider variety of products and materials.

18.2 Petroleum jelly

As the name indicates, petroleum jelly has for decades been produced from petroleum as a source material. In the United States, it was originally extracted from the crude oil found in Titusville, Pennsylvania, and the first oil wells. Since then, it has become a product produced on a large scale, routinely for individuals as end users.

https://doi.org/10.1515/9783111330259-018

Table 18.1: Biotech cosmetic companies.

Company name	Product trade name, example	Function	Website
Amyris	Squalene	Emollient	www.amyris.com
Contipro	Hysilk, hyaluronic acid	Antiaging	https://contipro.com
Ginkgo Bioworks	Rose oil	Fragrance	https://www.ginkgobioworks.com
Honest Co.		Body lotions, hand sanitizers	https://www.honest.com/
Lallemand	Probiotics	Yeast extracts	http://www.lallemand.com/
Natura Cosmetics		Body, skin care	https://www.naturabrasil.fr/en
Sederma	Sebuless	Oily skin, antiaging	https://www.crodapersonalcare.com/en-gb/our-brands/sederma
Solazyme		Microalgae oils	http://www.solazyme.com/
Sollice Biotech	Collamung	Skin soothing	https://www.sollicebiotech.com/en/
Twist Bioscience		DNA syntheses for other companies	twistbioscience.com

The petroleum jelly that consumers know as Vaseline®, a Unilever product, is a mixture of several hydrocarbons, in a proprietary formula. It is known to contain some amounts of white petrolatum and soft paraffin or paraffin waxes, and to be slippery to the touch. It has for years been marketed as a healing topical ointment.

In the recent past, there have been serious efforts to find alternatives to this product, not because of any danger, but because it is recognized that consumers would like products made from renewable, sustainable sources. Several companies now market personal care products that advertise different types of natural butters and essential oils which can heal chapped or rough skin. Examples of materials that are used in such products include cocoa butter, shea butter, mango butter, coconut oil, jojoba oil, and sweet almond oil. Table 18.2 is a nonexhaustive list of such firms.

Note that companies who manufacture nonpetroleum-based skin care products often promote the natural ingredients of their product right in the name. As well, this emphasis can be noted on the product packaging. An example is shown in Figure 18.1: a sample of Jao Brand goē oil that indicates its ingredients are all natural.

Table 18.2: Personal care companies promoting all natural products.

Product name	Company name	Website	Comments
Vegetable glycerin organic skin care	Aura Cacia	www.auracacia.com/	
goē oil	Jao Brand	jaobrand.com	Blends of plant-based oils and butters
Lip and skin balm	RMS Beauty	www.rmsbeauty.com	Blend of oils
Waxelene soothing botanical jelly	Waxelene	www.waxelene.com	According to the website: "all organic food grade ingredients"

Figure 18.1: Nonpetroleum-based body oil product.

18.3 Mineral oil

As the name implies, mineral oil is another organic material originally produced from crude oil, and found to be useful as a lubricant. What the name does not imply is that the term has been used, and marketed, for several different liquids, all of which include a mixture of higher molecular weight, nonpolymeric hydrocarbons. In the past, it was only after what may be called profitable amounts of it were isolated did its uses expand to cosmetics.

Mineral oil is used in numerous cosmetics, including infant skin care lotions and other skin creams. There is also a food-grade mineral oil. But because of its source, petroleum, there have again been some efforts to find other sources for it, or other products that can substitute for it in end-user applications. The results have been several products that are available in retail stores with names including terms that imply a biological source for their product, such as "Bio Renew." Perhaps obviously, such products work, but they are not necessarily made with mineral oil.

18.4 Personal care mixtures

Personal care products are often mixtures of a variety of ingredients, the aim of the mixture being to create a marketable product that has specific, desired properties. An obvious example is a mixture of materials that makes a skin cream which helps moisturize, color, or rehydrate the skin, while having good adhesion, and yet that does not feel greasy to the touch. While there is a wide variety of small molecules that can produce such a product (or any other personal care product), we will focus on two. Propylene glycol and butylene glycol have both been used in personal care product mixtures, and have both been examined for sourcing that is not petrochemically based.

18.4.1 Propylene glycol and 1,3-propanediol

This three-carbon molecule, propylene glycol, has traditionally been made from a petroleum base, and been used as a lubricant. Its Lewis structure is shown in Figure 18.2. It is a viscous liquid, and functions as a humectant in products such as hand sanitizers. Such sanitizers are often alcohol-based, and a humectant – a moisturizer – is needed in them to keep the skin from becoming too dry.

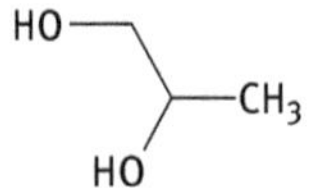

HO —⟍
     ⟩—CH$_3$
HO ╱

Figure 18.2: Structure of propylene glycol.

The material Susterra®, produced by DuPont, is a 1,3-propanediol alternative that is marketed as a "petroleum-free diol [15]." As well, another DuPont product, Zumea®, is marketed as a petroleum-free propanediol for cosmetic and food use. At the DuPont-

Tate-and-Lyle website, the company states: ". . .Zumea® propanediol is the multifunctional, preservative boosting humectant and ingredient that delivers high performance in a variety of consumer applications, from cosmetics and personal care to food, flavor, and pharmaceuticals, and laundry and household cleaning [15]." Both products are part of their "Biopreferred®" program. It is not mentioned specifically, but when possible the vegetable source for such chemicals is corn, and the glucose that it contains.

Despite detailed information on the safety and profile of uses of Susterra® and Zumea® at the website, the company does not release what organism produces their products, or how they are purified. While this is unfortunate for researchers, it is understandable when one considers the overall value of such products. As far back as 2001, the OECD mentions in its publication, "The Application of Biotechnology to Industrial Sustainability," that a pilot plant would be producing 90,000 kg of 1,3-propanediol per year [16].

18.4.2 Butylene glycol

Used as a solvent, as well as to decrease viscosity in some personal care mixtures, butylene diol has also been produced extensively from petrochemical sources. The isomer most used as a solvent is 1,3-butylene glycol, shown in Figure 18.3.

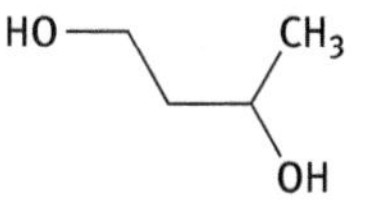

Figure 18.3: Butylene glycol.

Since the 1980s, however, it has been found that one isomer of it, 2,3-butylene glycol, can be produced from cheese whey. The microbial source can be *Bacillus polymyxa*, although more than one organism has been found to be effective in this production [17]. Once again, the use of an organism like *B. polymyxa* on any large, industrial scale is governed not only by how effective it is, but by how expensive or inexpensive it is when compared to traditional methods of production.

18.5 Recycling and reuse

The materials that have been discussed in this chapter are predominantly made into end-user materials, and thus are neither recycled nor reused. This has become an area of concern at least in the past decade, as these molecules are among many that survive and do not degrade in water purification plants. This in turn means that they are passed through such plants, and end up in rivers, lakes, and ultimately oceans. Efforts to adapt and build updated water treatment plants are only now beginning. The cost involved in such updating generally means that any such processes will require years to bring to fruition.

References

[1]   Independent Cosmetic Manufacturers and Distributors. Website. (Accessed 25 April 2024 as: https:// cosmoprofnorthamerica.com).

[2]   Personal Care Products Council. Website. (Accessed 25 April 2024, as: https://www.personalcarecoun cil.org/).

[3]   American Cosmetic Manufacturers Association. Website. (Accessed 25 April 2024, as: https://www. acma.us).

[4]   Society of Cosmetic Chemists. Website. (Accessed 25 April 2024, as: https://www.scconline.org/).

[5]   Cosmetics Alliance Canada. Website. (Accessed 25 April 2024, as: https://www.cosmeticsalliance.ca/).

[6]   Cosmetics Europe, the Personal Care Association. Website. (Accessed 25 April 2024, as: https://www. cosmeticseurope.eu/).

[7]   Colipa, the European Cosmetics and Perfumery Association. Website. (Accessed 25 April 2024, as: https://www.eesc.europa.eu).

[8]   Australian Society of Cosmetic Chemists. Website. (Accessed 25 April 2024, as: https://ascc.com.au/).

[9]   The Israeli Association of Cosmetics Manufacturers. Website. (Accessed 25 April 2024, as: cosmeticindex.com/).

[10] Cosmetic Toiletry & Fragrance Association of South Africa. CTFA. Website. (Accessed 25 April 2024, as: https://ctfa.co.za).

[11] The Cosmetic, Toiletry and Perfumery Association (CTPA). Website. (Accessed 25 April 2024, as: www.ctpa.org.uk).

[12] Cosmetics, Design-Europe.com. Website. (Accessed 25 April 2024, as: https://www.cosmeticsdesign-europe.com).

[13] ID Bio. Website. (Accessed 25 April 2024, as: https://www.cosmeticsandtoiletries.com).

[14] Biotechnology in Cosmetics: Concepts, Tools and Techniques, 2007, ISBN: 978-1932633245.

[15] Dupont, Tate and Lyle. Website. (Accessed 25 April 2024, as: https://www.tateandlyle.com).

[16] OECD. The Application of Biotechnology to Industrial Sustainability, 2001. (Accessed 25 April 2024, as: https://www.oecd.org/biotech).

[17] Speckman, R.A. and Collins, E.B. Microbial production of 2,3-butylene glycol from cheese, whey. Applied and Environmental Microbiology, May 1982, 1216–1218. Downloadable as: aem.asm.org/ content/aem/43/5/1216.full.pdf.

19 Biotechnology in recycling

19.1 Introduction

Recycling of materials, usually of metals, has a history that goes back more than a century, at least in terms of large-scale recycling of metals such as iron, brass, bronze, and tin. Usually the recycling in question was related to a war effort. In the First World War, metal scrap was collected for remelting and recasting into some usable item. Likewise, the Second World War saw numerous metal scrap drives to aid in the war effort [1]. While this is essentially the roots of industrial-scale recycling, it does not involve any biotechnological component.

In the decade immediately following the conclusion of the Second World War, a large number of new materials were developed, scaled up in production, and moved to the marketplace in an enormous array of consumer goods. Many of these novel materials are plastics. The ubiquity of plastics in our lives today has its origin in this post-World War II surge in development and production. Curiously, these plastics were originally designed to be extremely robust – to last "forever" – and not to degrade. The first serious thought into the problems related to their long life, and lack of what can be called natural degradation, came in the 1960s with the then nascent environmental movement.

Recycling today incorporates numerous materials, although metal, glass, plastics with resin identification codes (RICs) of 1–6 (discussed in Chapter 12), and paper are the largest volume materials. These four materials are recycled in large enough amounts that entire businesses exist that can be called recycling operations. As well, recycling as an industry is large enough now that there are numerous national and regional organizations devoted to it [2–17]. As recycling has become more widespread, these organizations have grown and evolved to include any applicable biotech operations that are effective, economically viable processes [18–23]. It has become common in the past decade to see trash receptacles in public parks, zoos, and other places where trash can be segregated, as shown in Figures 19.1 and 19.2. This helps ensure that noncompostable materials are sent for recycling, while compostable materials are disposed of or recycled separately.

It is noteworthy that in Figure 19.1 the containers are not simply labeled for different types of refuse; they are used as placards to help educate the public about the possibilities of recycling and landfilling.

Likewise, containers like those in Figure 19.2 are designed for larger items to be recycled without mixing their contents with compostable materials that may be recycled through some microbial means. But instructions can be placed on them indicating what is not permitted in such containers.

https://doi.org/10.1515/9783111330259-019

Figure 19.1: Public trash receptacle with separate bins.

Figure 19.2: Recycling receptacle for large items.

19.2 Microbes in recycling

Some form of microbial recycling of plastics and other long-lasting consumer end-use items remains in its infancy, but the past decade has seen significant strides in the use of enzymes and microbes to break down materials [1–7].

A much more established use of biotechnology in the field of recycling and reuse is the microbial digestion of organic materials, such as the solid components of wastewater and sludge, which has become an established means of dealing with such waste products.

The process of wastewater reclamation can be broken down into a few broad steps. They are:

1. Primary treatment. This stage removes macroscopic debris from incoming water.
2. Secondary stage. Often in this phase, some microbial or algal growth occurs, in which the organic material in the wastewater is the food source. After the growth,

the water can be filtered through sand beds or some other media, leaving the algae to dry and die, then later to be collected. Alternatively, this step can be run in a digester – a bioreactor if microbes are used – to help concentrate the product.

3. Antibacterial shock. This third phase is the point at which chlorine or some other antibacterial agent is added to the water. This does not need to be done if the water is to be discharged directly back to some natural water body, be it a river, lake, or ocean.

Beyond this, there is often a dry sludge residue to deal with. This can be used as fertilizer, or mixed with other fertilizers, for further use. Thus, as a whole, this process uses biological organisms to clean water and produce a residue, a sludge, that can be profitably used. Figure 19.3 illustrates this diagrammatically in some detail.

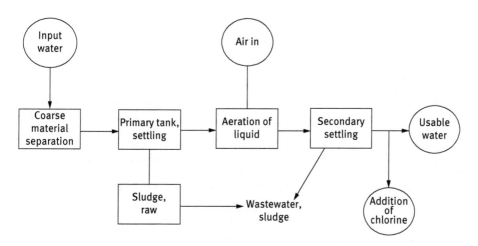

Figure 19.3: Flow diagram of wastewater treatment.

Note in Figure 19.3 the primary and secondary settling operations, as well as the aeration of the material (not used in all operations). These are the major steps in which microbial degradation can take place.

19.3 Recycling of plastics and biobased plastic materials

19.3.1 Plastics recycling

The production of plastics has become an enormous part of modern industry, and yet the end of their usable life is an area that is still in need of development. In other

words, while plastics are in theory recyclable an infinite number of times, the reality of the use of plastics is that most are not. Many plastic objects are used just once, then discarded. The reasons recycling is not universally practiced are many, but fall broadly into the following categories:
1. Sorting different plastics consumes time and can be expensive,
2. Only certain types of plastics are easily recyclable,
3. Recycling involves an energy input, which is an expense,
4. Recycled plastics are often of lower quality than that from which they were made [24].

Traditional forms of plastics recycling do not require a biotechnological step in their process, but rather involve remelting and reforming the material. In the past few years, however, attempts have been made to use bacteria, fungi, or other microorganisms to degrade plastics, often those found in landfills or at sea. The European Union has funded a project called BIOCLEAN, which seeks to determine what types of plastic can be degraded using a wide variety of living organisms. Their final report states in part:

> Microorganisms, i.e. bacteria (both anaerobic and aerobic) and fungi, but also enzymes produced by them, were thus selected as candidates for tailored polymer/plastic biodegradation screenings. Numerous candidate organisms were isolated from plastic wastes (hundreds of microbes obtained including aerobic and anaerobic bacteria, terrestrial and marine fungi, actinomycetes and anaerobic consortia from marine and terrestrial environments) in addition to those which were obtained from public or private collections (45 bacterial and 56 fungal strains).[25]

The efforts of BIOCLEAN were to examine the possible biotech-based clean-up of plastics such as polystyrene, polypropylene, polyethylene, and polyvinyl chloride. These and polyethylterephthalic ester (PETE) are the plastics produced on the largest scales, and thus those that are most often discarded into landfills or open waters [25, 26].

19.3.2 Biosourced plastics recycling

The obvious advantage of plastics made from biosourced monomers is that they represent a sustainable way to make these materials. By decoupling plastics from petroleum sources, it appears that they can be made indefinitely, at least in theory.

Interestingly, since biobased plastic materials differ only from their synthetic, petroleum-based counterparts in what the starting materials are, and not in what the final polymeric materials are (including their extensive covalent bonding), the process of their reuse, recycling, or degradation is the same for both. Perhaps obviously, this has been a continuing problem in moving to a position of a completely green plastics cycle. Current attempts at recycling such materials by breaking them down with a proprietary mix of enzymes have been announced by the European firm Carbios, which claims it can now break down both PETE and polylactic acid (PLA).

One means whereby biosourced plastics can be either degraded or recycled is to focus on plastics that were manufactured with some form of degradation in mind. The just-mentioned PLA, discussed in Chapter 12, is one such example. PLA is sourced from corn, and does degrade in composting conditions in approximately 6 months, depending on the amount of water to which it is exposed.

References

[1]   Power Recycling: A Division of Power Pallet. Website. (Accessed 25 April 2024, as: https://thinkpower recycling.com/).
[2]   National Waste & Recycling Association. Website. (Accessed 25 April 2024, as: https://wasterecycling. org/).
[3]   Solid Waste Association of North America. Website. (Accessed 25 April 2024, as: https://swana.org/).
[4]   National Recycling Coalition. Website. (Accessed 25 April 2024, as: https://nrcrecycles.org/).
[5]   Canadian Association of Recycling Industries. Website. (Accessed 25 April 2024, as: https://cari-acir. org/).
[6]   CIAC. Chemistry Industry Association of Canada, Plastics Division. Website. (Accessed 25 April 2024, as: https://canadianchemistry.ca).
[7]   Ontario Waste Management Association. Website. (Accessed 25 April 2024, as: https://www.owma. org/cpages/home).
[8]   Bluewater Recycling Association. Website. (Accessed 25 April 2024, as: https://www.bra.org/).
[9]   EuRIC – Advocating Recycling in Europe. Website. (Accessed 25 April 2024, as: https://www.euric-aisbl.eu/).
[10] European Association of Plastics recycling. Website. (Accessed 25 April 2024, as: http://www.epro-plasticsrecycling.org/).
[11] European Recovered Paper Association. Website. (Accessed 25 April 2024, as: https://euric.org).
[12] European Electronic Recyclers Association. Website. (Accessed 25 April 2024, as: https://www.eera-recyclers.com/).
[13] WMRR. Waste Management and Resource Recovery Association of Australia. Website. (Accessed 25 April 2024, as: https://www.wmrr.asn.au).
[14] Australian Council of Recycling. Website. (Accessed 25 April 2024, as: https://www.acor.org.au/).
[15] National Waste & Recycling Industry Council. Website. (Accessed 25 April 2024, as: https://www. nwric.com.au/).
[16] Australian Organics Recycling Association. Website. (Accessed 25 April 2024, as: https://www.aora. org.au/).
[17] Recycling Association of South Africa. Website. (Accessed 25 April 2024, as: https://recyclingassocia tion.wixsite.com/).
[18] Lady Green Recycling. Website. (Accessed 25 April 2024, as: www.ladygreenmiami.com).

[19]   Green Recycling Co. Website. (Accessed 25 April 2024, as: www.greenrecyclingco.com).

[20]   Xtreme Electronic Recycling. Website. (Accessed 25 April 2024, as: https://thekingofrecycling.com).

[21]   Generation Green Recycling. Website. (Accessed 25 April 2024, as: https://generationgreenrecy
        cling.com).

[22]   E-Waste, All Green Recycling LLC. Website. (Accessed 25 April 2024, as: https://www.all-green.com).

[23]   Sardon, H. and Dove, A.. Plastics recycling with a difference. Science, 360(6387), 380–381.

[24]   European Biotechnology, Life Science and Industry Magazine. (Downloadable as: https://european-
        biotechnology.com/the-mag/issues/issue/biotech-for-breaking-down-a-sea-of-waste.html).

[25]   BIOCLEAN Report Summary. (Downloadable as: https://cordis.europa.eu/result/rcn/177845_
        en.html).

[26]   Waste360.com. Website. (Accessed 25 April 2024, as: https://www.waste360.com).

20 Biotechnology companies

We mentioned in the first chapter of this book that the boundaries of the field of biotechnology are not clear cut; and indeed, they remain in flux as new techniques and applications are discovered. Table 20.1 is a list of biotechnology companies that, although up to date as this book is published, can never be considered a complete and total list, simply because the field is changing so quickly. The list is currently dominated by pharmaceutical and healthcare companies, but this is because many of the prescription and over-the-counter drugs on the market today are complex molecules that would be prohibitively expensive to sell if they were made through traditional, synthetic, organic chemical transformations.

Looked at from a different perspective, certain companies will probably never be on a list such as that in Table 20.1 because, even though they use some biotechnological means to produce a product or material, they do not consider themselves a biotech company. For example, as biotechnological techniques and steps are employed in increasingly wide areas, such as textile manufacturing, wood and paper processing, leather manufacturing, rubber production, and ore refining, such companies may need to employ scientists and engineers who are proficient in the use of specific biotech techniques. Yet those companies will still not consider themselves biotech companies, since they use other, often traditional, means to produce commodities or end-use items – or they combine biotechnological steps with older, traditional chemical transformations to manufacture or produce some final product or material.

Table 20.1: Index of biotechnology companies.

Abbott Laboratories	http://www.abbott.com/
AbbVie	https://www.abbvie.com/
Alexion	http://www.alexion.com/
Allergan	https://www.allergan.com/home.aspx
Amgen	http://biotechnology.amgen.com/biotechnology-explained.html
AstraZeneca	https://www.astrazeneca.com/
Baxter	https://www.baxter.com/
Bayer	https://www.bayer.com/
Biogen	https://www.biogen.com/en_us/home.html
Bristol-Meyers Squibb	https://www.bms.com/

https://doi.org/10.1515/9783111330259-020

Table 20.1 (continued)

Celgene	http://www.celgene.com/
Eli Lilly &Co	https://www.lilly.com/
Gilead Sciences	http://www.gilead.com/
GlaxoSmithKline	https://www.gsk.com/
Johnson & Johnson	https://www.jnj.com/
Merck & Co.	http://www.merck.com/index.html
Novartis	https://www.novartis.com/tags/biotechnology
Novo Nordisk	http://www.novonordisk-us.com/
Pfizer	http://www.pfizer.com/
Regeneron Pharmaceuticals, Inc.	https://www.regeneron.com/
Roche	http://www.roche.com/
Sanofi	https://www.sanofi.us/en/
Shire Pharmaceuticals	https://www.shire.com/
Stryker Corporation	https://www.stryker.com/us/en/index.html
Teva Pharmaceutical	http://www.tevapharm.com/

Index

https://doi.org/10.1515/9783111330259-021